A PRACTICAL GUIDE TO INSPECTING

EXTERIORS

By Roy Newcomer

CONTENTS

INTRODUCTION

My background includes many years in construction and several more as the owner of a Century 21 real estate franchise. In 1989, I started a home inspection company that grew larger than I ever imagined it could. Training my own staff of inspectors to the highest inspection standards led to my teaching home inspection seminars across the country and developing study courses, books, and videos for home inspectors. The American Home Inspectors Training Institute was founded as a result of my desire to share this experience and knowledge in home inspection.

The *Practical Guide to Inspecting* series is intended for both beginning and experienced home inspectors. So if you're studying home inspection for the first time or are using the materials as a refresher, these guides should be of assistance to you.

I've created these guides to include all aspects of home inspection. Not only a broad technical background in home systems, but the other things you need to know in order to perform a *good* inspection of those systems. They lay out technical information, guidelines for the inspection, how-to instructions for inspecting system components, and the defects, deficiencies, and problems you'll be looking for during the inspection. I've also included some advice on how to report your findings to the home inspection customer.

I've been a member of several professional organizations for a number of years, including ASHI® (American Society of Home Inspectors), NAHI™ (National Association of Home Inspectors), and CREIA® (California Real Estate Inspection Association). I am a great supporter of those organizations' quest to promote excellence in home inspection.

I encourage you to follow the standards of the organization to which you might belong, or any state regulation that might take precedent over the standards used here. Use the standards in this book as a general guide for study and apply the standard or state regulation that applies to you.

The inspection guidelines presented in the Practical Guides are an attempt to meet or exceed standards and regulations as they exist at the revision date of the guides.

There's a lot to learn about home inspection. For beginning inspectors, there are some *hands-on exercises* in this guide that should be done. I'm a great believer in learning by doing, and I hope you'll try them. There are also some of my *personal inspection stories* to let you know what it's really like out there.

The *inspection photos* referenced in this text can also be found on www.ahit.com/photos. You'll read the story about each one as you go along. Be sure to watch for my *Don't Ever Miss* lists. I've included them to alert home inspectors to report those defects (if found during the inspection) in the inspection report. If missed, these items are often the cause for lawsuits later. Finally, to help you see how you're doing as you study this guide, I've included some *worksheets*. The answers are given for each one for self checking. Give them a try. Checking yourself can help you lock important information in your mind. There's also a *final exam* that you can complete and send in to us. Many organizations and states have approved this book for continuing education credits. Submit the exam with the required fee if you need these credits.

In total, the *Practial Guide to Inspecting* series covers all aspects of the general home inspection. Each guide covers a major aspect of the inspection, as their titles show:

Electrical
Exteriors
Heating and Cooling
Interiors, Insulation, Ventilation
Plumbing
Roofs
Structure

If you are interested in other titles in the series, please call us at the American Home Inspectors Training Institute to order them. Call toll free at 1-800-441-9411.

Roy Newcomer

INSPECTING
EXTERIORS

Chapter One

THE EXTERIOR INSPECTION

Inspection Guidelines and Overview

These are the standards of practice that govern the inspection of the exterior components of the property.

Guide Note

Pages 1 to 4 lay out the content and scope of the exterior inspection, which includes the exterior components of the property, excluding the roof. It's an overview of the inspection, including what to observe, what to describe, and what specific actions to take during the inspection.

Exterior Systems	
OBJECTIVE	To identify major deficiencies in the condition of exterior components of the property, including attached structures.
OBSERVATION	<u>Required to inspect and report:</u> • Wall claddings, flashings, and trim • Entryway doors and representative number of windows • Garage door(s) and operator(s) • Decks, balconies, stoops, steps, areaways, and porches including railings • Eaves, soffits, and fascias • Vegetation, grading, drainage, driveways, patios, walkways, and retaining walls <u>Not required to observe:</u> • Storm windows, storm doors, screening, shutters, awnings, and similar seasonal accessories • Fences • Safety glazing (depends on standards) • Garage door operator remote control transmitters • Geological conditions • Soil conditions • Recreational facilities • Outbuildings other than garages and carports
ACTION	<u>Required to:</u> • Operate all exterior doors and a representative number of windows. • Report whether or not any garage door operator will automatically reverse or stop when meeting reasonable resistance during closing.

- Wall claddings
- Trim
- Doors and windows
- Attached structures
- Drives, walkways, patios
- Grading and vegetation
- The garage
- Retaining walls

Definitions

Wall cladding, or siding, is a covering for the exterior of the house that protects the framework of the structure.

The eave, or overhang, is the lower portion of the roof that extends beyond the outer wall. It is made up of the fascia, which is the outer board laid vertical at the edge of the eave, and the soffit, which is the underside of the eave.

Not every detail of what is to be inspected and what is to be reported is listed in these standards. However, they do present a good overview of the exterior inspection.

Note that the standards of practice for inspecting the exterior of a house explicitly list a number of items that the home inspector is *not* required to observe and report on. For clarity, the standards define the inspection by listing both what to do and what not to do. We are going to suggest a few deviations from the standards in that we like to inspect and report on the condition of storm windows and screens. We suggest the home inspector identify the materials used and report on their general condition.

Here is an overview of the exterior inspection:

- **Wall claddings:** The home inspector examines the wall cladding, or siding, covering the exterior of the house. The inspector is required to describe the **materials** used as siding. Examples of siding materials are clapboard, wood shingles and shakes, asbestos shingles, brick, stone, aluminum, and vinyl.

  The home inspector is also required to report on the **condition** of the siding and report any **defects** found. Sidings can warp, crack, twist, leak, and come loose from the walls. The inspector looks for deterioration of materials such as rotting wood and spalling brick, peeling paint, cracking stucco, and so on. Missing moldings and caulking in need of repair are also reported.

- **Trim:** The home inspector examines the trim on the house, including the eaves, soffits and fascias. The materials used for trim are reported as well as their condition. The home inspector pays close attention to wood rot. The tip of a screwdriver, ice pick, or knife should be used to probe the trim when deterioration is visible or suspected.

- **Doors and windows:** The inspector is required to inspect and **operate** all entryway doors and a **representative number** of windows. A representative number of windows means enough to have a fairly accurate idea of their condition. The inspector identifies the materials used in window construction and reports on the condition of these materials. Particular attention is paid to whether wood rot is present, caulking needs repair, and a new paint job is needed. All doors should be opened and inspected for ease of operation and condition. The home inspector is not typically required to inspect or report on **safety glazing**.

NOTE: We also advise the home inspector to inspect the **storm windows and screens**. Notice that the standards of practice state that the home inspector is not required to observe storm windows, storm doors, screening, shutters, awnings, walkways, and similar seasonal accessories. We find an exception in storm windows and screens because it's an easy and quick process to observe and report torn screens and cracked or missing storms.

- **Attached structures:** The home inspector carefully inspects all attached structures such as balconies, decks, porches, stoops, and stairs. The structural integrity of these attachments is one issue the inspector pays attention to, examining support posts and beams, flooring, and railings. Since these structures are exposed to the elements, care is taken to probe them for wood rot and deterioration. **Safety features** of steps and stoops are examined.

- **Drives, walkways, and patios:** When the home inspector inspects these areas, attention is paid to their relation to the house itself — whether drives, sidewalks, or patios slope toward the house or show gaps at the junction of the house and are likely to send water towards the structure. They're also examined for materials and condition. Cracks and gaps in these surfaces are noted as trip hazards. The general home inspection does not include **recreational facilities** such as barbecues, fire pits, tennis courts or pools.

- **Grading and vegetation:** The home inspector is not a landscape designer passing judgment on the pleasing effects of grading and vegetation. But the inspector determines if the slope of the land around the house and trees and bushes near the house are having a detrimental effect on the foundation. The inspector does not have to inspect or report on **soil** or other **geological conditions**.

- **The garage:** The garage is also described and inspected. The outside siding and trim are inspected along with the rest of the house and reported separately with other garage

Guide Note

Other items are inspected while the inspector is outside the house. They're covered in another one of our guides. The roof covering and the roof drainage system are presented in A Practical Guide to Inspecting Roofs. The exterior electrical features that must be inspected are introduced in A Practical Guide to Inspecting Electrical. The air conditioner compressor/condenser unit is also inspected. That's presented in A Practical Guide to Inspecting Heating and Cooling.

features. Attention is paid to windows and doors, especially the overhead garage door. The home inspector is required to test the operation of the **safety reverse feature(s)** of the automatic door and report its malfunction as a safety hazard. The home inspector does not have to test the remote control garage door opener. Inside the garage, the home inspector identifies the floor materials and reports its condition.

The home inspector also inspects the structural integrity of the garage and reports on its general condition.

Other **outbuildings** such as gardening sheds and barns are not considered to be a part of the general home inspection.

- **Retaining walls:** The home inspector observes the condition of retaining walls that are close to the house and may have an effect on the structural integrity of the house. For example, a retaining wall holding back soil next to the house would be inspected. A wall at the back of the property, which has no effect on the structure whatsoever, would not be inspected. The home inspector inspects the wall for movement and stability, checking the wall's ability to discharge water.

Chapter Two

INSPECTING THE SIDING

An important aspect of the exterior inspection is the inspection of the wall cladding or siding. The home inspector identifies the **type** of wall cladding used and reports on its **condition**. Siding should be inspected from all sides of the home, not just selected views. The inspection of the siding includes:

- Condition of siding materials
- Loose or missing components
- Siding fastenings
- Condition of the joints and interfaces with other materials
- Finish paints and stains
- Distance from the ground

Under the Siding

The purpose of the siding is to protect the framework and the interior of the structure from the elements. In some cases, the siding also contributes to bracing the structure and enhancing its rigidity. With solid masonry walls, of course, the exterior walls are also structural members of the structure.

In old balloon frame homes the siding may have stabilized the structure without the help of any **wall sheathing** beneath it. But more recent construction includes wall sheathing to brace the framework and provide some insulation and weather resistance.

In older platform frame homes wall sheathing was **tongue and groove planking** nailed on the diagonal to the wall studs. **Building paper** was applied over the planking to act as a water repellent and air barrier. It also acted as a lubricating layer between wood surfaces. The siding was nailed to the planking through the building paper. The cavities between the studs may or may not have been filled with insulation.

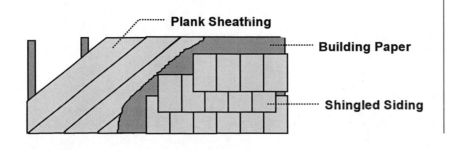

WALL SHEATHING

- Diagonal planking
- Plywood
- Fibrous sheathing

WATER AND AIR REPELLENT LAYER

- Building paper
- Non-woven plastic fabric
- Bead board
- Foil-coated glass fiber batting

Definitions

Wall sheathing is planking foam OSB or plywood sheets used to cover the wall framework of a structure.

When a surface is moisture permeable it allows moisture to pass through it.

Flashings are a type of sheet metal used at interfaces between building components to prevent water penetration. Flashings are used around chimneys and vents in the roof, above doors and windows in exterior walls, to protect the top of the foundation, and so on.

Most recent platform frame structures make use of OSB or **plywood sheathing** which serves to brace the home. If the house is well braced through some other bracing method, a **fibrous sheathing** may be used. But with a fibrous sheathing, the siding should add stiffness to the framing. Today, instead of building paper, it is more likely that a **non-woven plastic fabric** that comes in large rolls is used to protect against water and air penetration. Bead board or foil-coated glass fiber batting, combination air barriers and insulation, may also be used under the siding. In most climates, the stud cavities under the sheathing are filled with insulation.

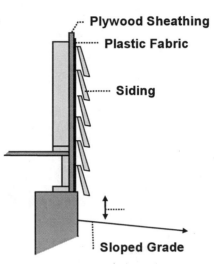

In general, the exterior layers of the structure should be **permeable**. That means that moisture in the air should be allowed to pass through and not be trapped in the walls. If a moisture deterrent layer is used, it should be inside the wall studs, not outside. Plank, plywood, and fibrous sheathing are permeable materials as are the building papers and fabrics used to repel water and air infiltration.

During one construction period, polystyrene foam boards were installed as insulation between the sheathing and the siding. This was unsatisfactory, especially for plank siding, because foam boards were moisture impermeable and moisture passing through the plank siding was trapped. If broken or split plank siding or pulled nails are found, an underlying layer of impermeable foam boards should be suspected.

Exterior walls can be made too air tight through excessive caulking. Condensation within the wall cannot escape to the outside, leaving the wall damp and subject to rot.

Flashings are used under the siding to prohibit water penetration. They are generally used at the interfaces between the siding and other building components where water could easily leak in. Flashings are made of a type of sheet metal.

One example of where flashings are found under the exterior

siding is at the top of the foundation to keep water from penetrating into the foundation wall. For example, with a brick veneer wall, a flashing runs over the foundation beneath the brick and up the wall behind the brick to provide protection.

Where the exterior wall of the house's upper story meets a roof, flashing should be used to prevent water from entering at this interface. Ideally, flashings should be used over the top of window and door trim, that projects out from the siding, unless positioned under an overhang.

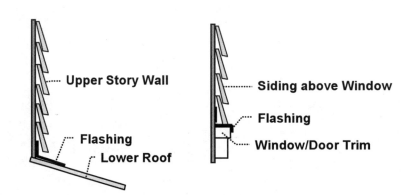

Upper Story Wall

Siding above Window

Flashing

Flashing

Lower Roof

Window/Door Trim

Wood Plank Siding

Wood plank siding can be installed either horizontally or vertically. These planks can be cut perfectly rectangular, tapered, or with special milled cuts.

Rectangular planks are perfect rectangular cuts which can be laid flat with their ends butted for a smooth exterior surface. These butted planks are usually installed vertically as siding. Vertical joints may be protected with **battens**. When rectangular boards are laid horizontally and overlapped, the style is called **clapboard**. This name comes from clamping or *clapping* the horizontal planks together with nails. Clapboard has a softer line if the planks are tapered or **beveled** instead of perfectly rectangular.

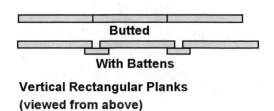

Butted

With Battens

Vertical Rectangular Planks (viewed from above)

Clapboard

Horizontal Planks (viewed from side)

WOOD SIDINGS

- Plank
- Clapboard
- Shingles and shakes
- Plywood Panels
- Composition board or hardboard

Battens are narrow strips of wood placed over joints in vertical wall siding. They serve to seal the joint.

Clapboard is overlapping, horizontal wood plank siding, made from either rectangular planks or beveled or tapered planks.

Various **milled planks**, as shown here, are used for plank siding, including tongue and groove siding which is not shown. **Shiplap** style planking is laid close enough that it appears to the eye to be butted.

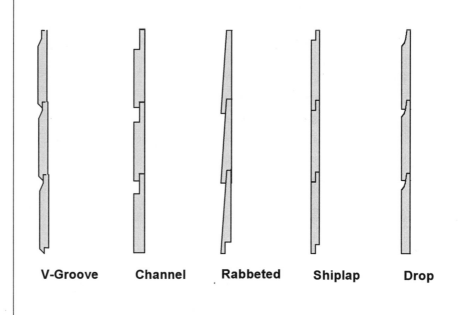

| V-Groove | Channel | Rabbeted | Shiplap | Drop |

The home inspector should carefully inspect wood plank siding for:

- **Fastenings:** Proper nailing plays an important role in the life of the siding. Too many nails can limit the wood siding's ability to expand and contract with moisture and temperature changes. Nails too close to the edges can result in splitting the wood. When face nailing is done as with clapboard, rabbeted bevel, and channel siding, nails should be set and puttied. Proper nailing decreases warping, checking, cupping, and the tendency of nails to pull out. Watch for the following problems with the nails themselves:

  — Nails too short and not holding
  — Nails too long and sticking out
  — Nails too thick that are splitting the siding
  — Corroded nails that are discoloring or disintegrating the wood
  — Unputtied nails that leave rusty streaks on paint
  — Nails that don't seem to grip

  The last item in the list — nails that don't seem to grip —

can signal another problem in vertical plank siding. The nailing base may not be proper. When vertical siding is installed, there should be horizontal blocking between the studs every 24" for the vertical planks to be nailed into. The wall sheathing is usually not strong enough or thick enough to hold the nails.

- **Deterioration and finishes:** Wood siding should be painted or stained to form a protective seal on the wood. Even cedar and redwood are helped with staining. The coating should be free of peeling, blisters, bubbles, dirt, and worn areas. If not, the condition should be reported and exposed siding inspected for **wood rot**. Wood rot can be hidden under new paint surfaces. Sometimes, you can see slight surface irregularities in the paint that are characteristic of hidden decay. In any case, if you suspect wood rot, you should probe the siding to make sure.

Blistering or peeling paint in a localized area should be explored for its cause. It may be from an obvious cause such as a leaking gutter or water splashing against the house at a corner. It can also be an indication that water or excess moisture is penetrating into the siding from the interior of the house. An example might be where water from a roof leak is running down the rafters and into the wall.

Broken or split plank siding and pulled out nails can be caused by improper nailing (see page 8). But another cause can be the presence of polystyrene foam boards behind the siding which causes moisture to be trapped in the siding and causes deterioration (see page 6). When plank siding absorbs and retains too much moisture, **cupping** and **checking** of the planks can result. Any cracks appearing in wood allow additional moisture to enter the wall.

Cupping or Warping

Checking or Cracking

Distortions: Warping and buckling of plank siding can be an indication of deterioration of the siding or a problem with the framework. You might see humps, bulges, and low places in the siding that can be the result of the

(see page 8)
(see page 6)

INSPECTING WOOD SIDING

- Nail problems
- Decay and wood rot
- Blistering and peeling paint
- Warping, buckling, and cupping
- Water penetration at interfaces
- Broken, loose or missing components
- Distance from the ground

Definitions

Cupping in wood plank siding is a warp across the grain of the board.

Checking in wood plank siding is a crack or split along the grain of the board as a result of cupping.

The best way to prepare yourself for inspecting wood and other types of siding is to get out there and take a look at some real homes. Before you focus on defects, concentrate on being able to identify the types of siding you see. Identification of the type of siding is an important part of the home inspection.

*Photo # 1 shows an example of **clapboard siding**. If you sight down this wall, you can see a bowing out of the lower courses of clapboard. It was a slight bow of about 2 1/2", but we investigated the source of this problem. We discovered that the bottom plate was pushed all the way out beyond the sill.*

movement of the frame or foundation. It could also be the result of warped or crooked studs pushing on the exterior siding or warped sheathing.

When vertical siding bulges out and nails are loose, this may indicate an absence of a proper nailing base. Distortions or displacements of planks make openings for water to penetrate behind the siding and wood rot can be expected.

#1 Clapboard siding

- **Distance from the ground:** Wood siding should end 6" to 8" above the soil — higher in areas where termites are a problem. Check the bottom edge of the siding to see if it touches the ground. Wood rot and insect problems will be present if it does. Take another look at **Photo #1**. This clapboard siding goes all the way to the ground and has no clearance from the soil at all. When we probed the lowest board around the house, evidence of wood rot was found.

- **Joints and interfaces:** Always inspect the joints in wood plank siding for water penetration and wood rot. Where battens are used to cover joints in vertical planking, check the top surface of the batten for deterioration. Problem areas can be where there are horizontal joints in vertical planking. Caulking may be used to seal joints between planks or cracks in the wood. If that is the case, be sure to inspect it to make sure it's in good condition.

Check interfaces at corners, where siding abuts a roof, and so on. Be sure these areas are water tight and free of deterioration.

- **Loose or missing components:** Get close enough to check the siding. Push against it to test its tightness to the framework. Wood plank siding should always be checked for any missing parts. Watch for any missing battens and report them.

Plywood Siding

Plywood sheets with weather resistant glue are used as exterior siding. They must be securely nailed in place with a sufficient number of nails in order to contribute to the stiffness of the framework. The backing for plywood siding should be sturdy. Where plywood sheets meet horizontally, there should be extra horizontal blocking between the studs to provide a proper nailing base.

Scarfed Joint

The outer ply of plywood siding can be disguised in a variety of ways. A common approach is to groove the outer ply to look like vertical planking and to sandblast it for a grainy appearance.

The home inspector should check the joints — both vertical and horizontal — for tightness. Horizontal joints are particularly vulnerable to water penetration and should have a flashing behind them or be **scarfed** rather than butted to prevent water from getting in. Vertical joints may have battens for extra protection that makes the siding look even more like vertical wood planking.

Exterior plywood can expand and contract at different rates than the framing, causing joints to be pulled apart from the movement. Nails can be pulled out and panels can actually fall off. Relief joints may be provided to counteract this problem.

Plywood siding has low permeability for water vapor and can absorb a great deal of moisture into the siding. The siding can easily warp if surface finishes are allowed to deteriorate.

For Beginning Inspectors

Make some stops at siding suppliers to see firsthand the variety of siding material available — in wood, aluminum, and vinyl. It would also be helpful to visit suppliers who carry stone and brick facings.

*Photo #2 shows **plywood siding (T-111)**. Here you can see the grooves made in the panels and the grainy surface, done to imitate wood plank siding. Note the area where the two vertical "planks" have pushed out. There is pressure on the plywood panel from warping or differing expansion and contraction of the framework behind the panel. The panel has nowhere to go as it expands. The decorative grooves in the outer ply, being weak points in the surface of the panel, can bend or break, looking even more like individual planks.*

Personal Note

"I want to emphasize the importance of inspecting the siding from <u>all sides</u> of the house. Don't inspect one wall and make assumptions about the others. Inspect the entire structure.

"One of my inspectors was inspecting a ranch-style home with extensions outward and expansion upward. The inspector walked around the entire house to review the siding and discovered that four different sidings had been used — concrete block, brick, wood, and vinyl. Turns out the owner was a construction laborer who had brought materials home from building sites, and then went nuts building additions.

"It took the inspector some time to thoroughly inspect the interfaces between these sidings."

Roy Newcomer

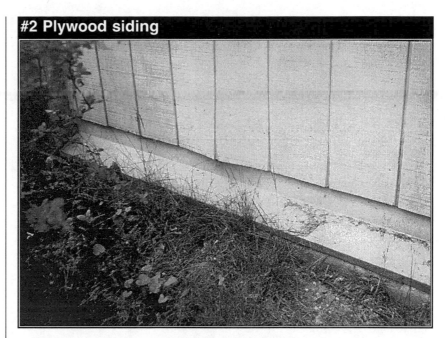

#2 Plywood siding

Plywood siding should be inspected as plank siding — finish, condition, distortion, joints, and distance from the ground, and, of course, for delamination.

Composition Board

Composition board or hardboard is made of **compressed wood fibers** with weather resistant binders. It can be made into planks or sheets for exterior siding purposes. However, boards and sheets made with wood *chips* or *ground* wood are usually intended for the inside of the house.

A sign that siding is composition board panels can be long runs of siding without joints, up to as much as 16' without interruption. But when made to imitate planks and installed horizontally, this fabricated siding can look like clapboard.

The home inspector should pay particular attention to the condition of composition board and hardboard, as both are likely to absorb moisture if not properly protected. Boards can expand and bow out, causing a wavy appearance in the wall. They may be swollen at the edges and bottom from moisture absorption. The boards can swell, warp, and disintegrate. Repainting of hardboard siding needs to be done more often than wood siding due to the material's lesser ability to hold paint. The home inspector may notice watery-looking, thinning, or fading paint on composition board and hardboard.

Wood Shingles

Wood shingles are usually redwood, cedar, or cypress. Shingles are rough or smooth sawn; **wood shakes** are handsplit. They can be installed in a variety of ways. When laid **single course**, each shingle is exposed to the weather (see below). With **double course** shingles or shakes, an undercourse lies completely behind and covered by a top course (see opposite page).

In one type of single coursing, the shingles can be laid 1/2 or 1/3 of their length to the weather, as seen here in the drawing on the left. This results in three layers of shingles.

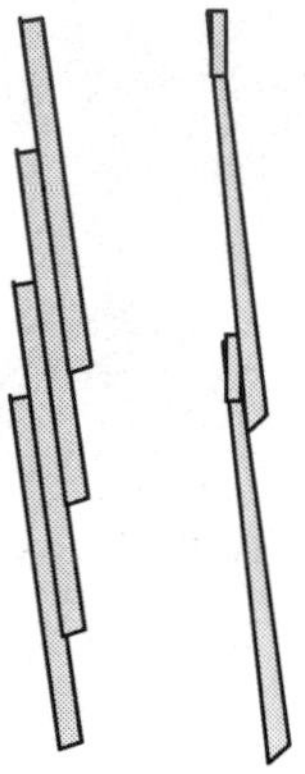

Another approach to single coursing is to use a tapered shingle (drawing at right), where the top end is very thin. This end is pinched under a **nailing strip**, and the butt edge of the next course is then nailed to the strip. A much longer portion of the shingle is exposed with this approach.

In double coursing, two layers of shingles are laid for each row. They may be even at the butt end or, as shown here, the top course may extend an inch or more beyond the undercourse. This is done for decorative reasons to enhance the shadow line at the bottom of each course.

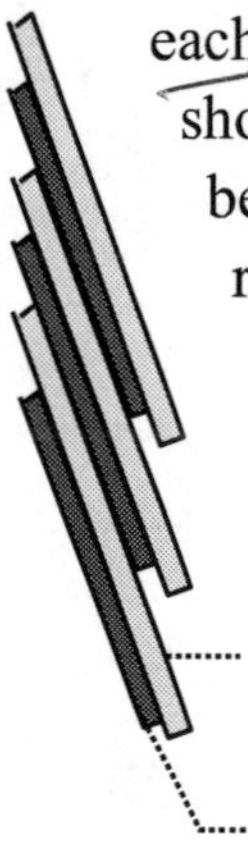

Top Course

Undercourse

Shingling can vary from rough, uneven looking patterns using different sizes of shakes all the way to very uniform sizes of shingles, which when painted, can almost look like clapboard.

Shingles with fancy cuts at the butt end, either pointed or rounded, was common in Victorian times. In the 1920's, a popular style was to lay alternating courses at different exposures for a more interesting pattern.

Cedar, redwood, and cypress will weather naturally, but different exposures to sunlight and moisture causes uneven aging and coloring. Discoloration is not a problem, although some homeowners may be concerned when it's uneven. But

Definitions

A shake is a type of wood shingle that is handsplit instead of being rough or smooth sawn.

A single course of shingles or shakes is where each course is exposed to the weather. In a double course of shingles or shakes, an undercourse, covered completely by a top course, is not exposed to the weather.

shingles and shakes may be stained for a uniform look and should be renewed periodically. And they can be painted, although once that's done; they'll always have to be painted.

The home inspector should be on the lookout for:

- **Warping** and **cracking** from age. Shingles also cup and split when they're not allowed to dry out.

- **Buckling** shingles caused by joints too tightly spaced, leaving no room for shingles to expand.

- Joints that are **not staggered** from course to course, and split shingles that create new joints directly below joints above. This creates a vertical alignment that promotes the flow of water behind the shingles.

- **Missing** shingles and shakes.

- **Loose** shingles or shakes. They should fit tightly. You shouldn't be able to lift them up.

- As with other sidings, **distortions** such as bowing or sagging of an exterior wall of shingles can indicate a problem with foundation settlement or warped framework.

*Photo #3 shows **wood shingles** on an upper floor of a house. For the most part, these shingles are in good shape. Some of them show some wear. In general, not until about 15% of shingles are in bad shape does the siding need to be replaced. The discoloration of the shingles near the top of the wall, although unsightly, is not a problem. What's wrong with this picture is the condition of the window and its effect on the shingles below it. The window sill has totally rotted away. We'll talk more about inspecting windows later in this guide.*

#3 Wood shingles

Metal Siding

Exterior metal siding may be made of either aluminum or steel. Neither type requires much maintenance, both come in smooth or embossed surfaces. Aluminum is the more popular metal siding these days. Steel is not used any more because of its tendency to rust in exposed areas.

Aluminum siding has a baked-on enamel surface that stands up over time. It can fade over the years, and some homeowners paint the siding. If the siding has been painted, then it will require repainting on a regular basis.

Aluminum siding wears well, although it can become dented. And it can be noisy. Because of expansion, it is not uncommon to hear it bang and pop when sunlight warms a side of the house.

Most of the problems with metal siding are the result of **defective installation procedures**. The siding should be securely and tightly fastened to the wall. When joined lengthwise, metal planks should be overlapped. The home inspector will want to check that these overlaps are big enough to prevent water penetration. The proper trim and moldings at corners and around doors and windows should be installed and

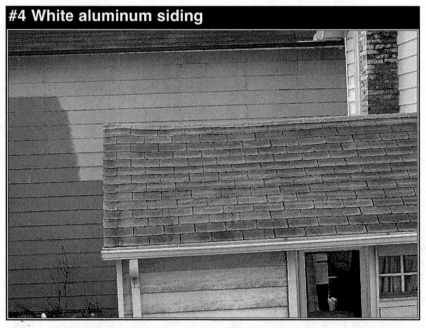

#4 White aluminum siding

Photo #4 *shows the back of a property covered with **white aluminum siding**. The smaller addition at the back of the house shows how dirty this siding can get. The home inspector can point out that this siding can be cleaned — sprayed occasionally with the hose and scrubbed down, or in more serious cases, cleaners can be purchased to clean it.*

Personal Note

"Photo #4 is interesting for another reason. I was actually inspecting the house next to the one with aluminum siding. But the houses were so close together, I couldn't get a look at this side of the brown house. I stepped into the yard next door to view it. And am I glad I did.

"The brown siding is clapboard and the paint job at this back corner hadn't been finished."

Roy Newcomer

securely fastened. At the corners, be sure each corner molding is overlapped by the one above. It's not unusual to have installers lap them with the lower molding overlapping the one above.

The home inspector might find aluminum siding installed with **no sheathing** under it. This is not a problem if the framework is properly braced. But if installers removed old siding *and* sheathing when installing replacement aluminum siding, the framework can be severely weakened.

Caulking is sometimes used with aluminum siding — often in the corners around windows and perhaps at overlaps or corners. The home inspector should inspect caulking and determine whether it's still doing its job.

Another thing to watch out for is the siding's inability to allow **moisture to escape** through the wall to the outside. Aluminum planking is waterproof, and that's good, but it should not be a moisture barrier. To keep it from becoming a moisture barrier, small holes appear along the bottom of each plank. If there aren't enough holes or if the holes are blocked, the siding can trap moisture inside the wall. The home inspector should check for these holes and look for rot behind the siding by examining the header joists in the basement and along the outer floor edge in the living areas.

Some local building codes require metal sidings to be **grounded**. There is a difference of opinion nationwide about whether grounding is needed. However, if your local codes require grounding, you should look for it.

Twisted, buckled, and bent aluminum siding may be an indication of structural problems. Always investigate further. Twisted aluminum siding on a house we inspected is an example. Upon investigating the foundation, we discovered serious settlement problems at the corner of the house.

Vinyl Siding

Siding made of polyvinyl chloride planks acts very much like aluminum siding. One difference is that vinyl siding can be easily damaged. It becomes brittle in temperatures at or below freezing and can break on impact. If inspecting under these conditions, be careful about putting your ladder against the siding or banging anything against it.

Again, most problems come from defective installation. Because vinyl siding has less strength than aluminum, sheathing should be present. Inspect vinyl siding for the same defects as you would metal sidings.

The home inspector can tell the difference between aluminum and vinyl siding by making a scratch test. Use a knife to make a small surface scratch at the end of a panel or joint. Vinyl siding is colored all the way through; aluminum siding has only surface color.

Asphalt Composition

Asphalt composition shingles used in siding are essentially the same as that used for roofing. They were popular in the 1930's through the 1950's, often placed over the original siding. This siding consists of strips of shingles laid in course with the higher courses overlapping the lower. Today, asphalt shingles come with an adhesive on the upper portion of the strip which bonds to the lower portion of the strip above. This adhesive doesn't always work as well as it should, and shingles can lift, curl, and be damaged when blown by the wind.

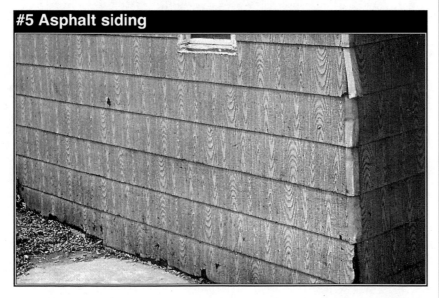

#5 Asphalt siding

*Photo #5 shows an old house with **asphalt siding**. The siding should be inspected for deterioration. Note the lower course of shingles, which should have at least 6" clearance from the ground. Here, siding has been damaged by water at ground level and where covered with soil. This siding cracks, and chunks fall off when exposed to this much wear. Note the corner molding out of place and moldings missing at the lower courses. These two conditions were reported. Other than those problems, the siding is in fairly decent condition — no cracking or deterioration, no lifting and curling, and joints well sealed.*

Asbestos Cement

Asbestos cement siding is a material made of asbestos and portland cement and is fireproof and weather resistant. It was in popular use in the 1950's. The asbestos present in the siding poses no health danger as long as the siding is in place. However, scraping the siding to clean it before painting can release asbestos fibers. Fibers can also be released when the siding is removed. The EPA requires that removed asbestos cement siding be treated as a hazardous material and disposed of in an approved landfill. Customers should be informed of this information whenever you find asbestos cement siding.

This siding material is durable. It is, however, brittle and can be cracked or broken on impact. (Be careful not to bang your tools against asbestos siding and break it yourself.) The color of the shingles can dull over the years, and the home inspector may find the siding painted. The finish paint should be inspected as other paint jobs.

When inspecting asbestos cement siding, report cracked, chipped, missing, loose, or displaced shingles. Check to see that nails are not loose or backed out. Report excessively soiled or stained patches and moss growth, which this siding seems to attract.

Stucco

Applying a stucco finish to a structure is a lot like plastering the interior walls. Stucco is a compound material made from sand, cement, and water. Today some stucco finishes are acrylic. With traditional stucco over wood framework, a wire mesh is attached to the sheathing and studs. Two or three coats of stucco of various mixtures are then applied to the mesh. Finally, the stucco top coat is applied in either a smooth or textured finish or to resemble stone with mortar joints and raised stone pieces.

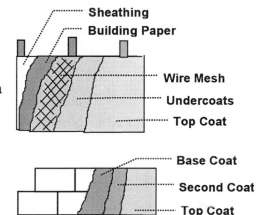

Stucco can be applied over masonry walls as well. In this case, the wall provides enough stiffening to bypass the wire mesh layer. One or two undercoats are applied before the top coat.

Stucco is water resistant and weatherproof, but is permeable to water vapor to let moisture escape from the wall. If water does penetrate the stucco finish through cracks or there is moisture buildup in the wall, stucco can deteriorate. Watch for defects in gutters and downspouts, flashings, and drip edges that cause water to leak down behind the stucco.

Inspect stucco walls carefully for **cracking** and try to find the cause of it. Cracks that go through all layers, cracks that hold

---

INSPECTING STUCCO

- Detachment from wall
- Water penetration
- Cracks that hold water
- Offset cracks and breaks
- Cracks that go through all layers

---

water, and offset cracks and breaks in the stucco should be reported to the customer and immediate repair suggested. Cracks can appear in stucco over wood framework when the **framing members shrink**, especially at the floor levels where there tends to be the most shrinkage. Vertical and horizontal cracking can take place when there is **foundation settlement**. Cracks in random directions can appear during the **curing** of the stucco. These types of cracks may be hairline cracks in the outer surface only or be deeper, involving the undercoats as well.

Stucco can become **detached** from the wall due to trapped moisture. Bulging and sagging can be a sign of it. Often, homeowners will paint stucco after repairs are made using an impermeable paint. Stucco paint should be a permeable masonry paint which allows moisture to escape. The home inspector can tap the wall with a screwdriver handle and listen for signs of detachment. Areas in the wall coming loose will make a dull thud when tapped. With acrylic stucco, which is applied over a polystyrene sheeting and a fiber mesh netting, push sharply against the wall with the heel of your hand to see if the surface is pulling away from its polystyrene base.

Stucco is sometimes combined with wood in Tudor or half-timbered exterior walls. In modern Tudors, there are painted wood trim pieces nailed to the sheathing and surrounding panels of stucco. When inspecting this type of exterior wall covering, pay special attention to wood rot in the wooden members, especially in the horizontal ones.

#6 Stucco siding

*Photo #6 shows a house with **stucco siding** which is in fairly good shape. However, close investigation at the corner on the left showed evidence of earlier repair work that is beginning to crack open again. This problem was originally caused by a defective downspout at the corner. We discovered that the downspout connection at the top was not tight and was still sending water cascading down the outside of the downspout. The owners had repaired the wall but not the cause of the problem.*

Solid Masonry Walls

The walls of a home can be made of solid masonry. Usually, there are two thicknesses of masonry. The interior wall may be left exposed, plastered over, or covered with drywall. The masonry wall can be made of such materials as brick, stone, and concrete block.

A **solid brick wall** can usually be identified by the **header rows**, where the brick is turned with its small end facing out. The header rows serve as ties to hold the bricks together. (Header rows are not present in brick veneer walls.) The thickness of brick walls has declined over the centuries from 20" thick to 16" to 12" and finally to 8" thick. Most brick homes built since the early 1970's are, in fact, brick veneer over a wood framework.

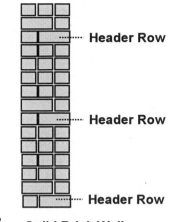

Solid Brick Wall

The solid masonry wall may be built of two materials — an inner and an outer layer. This is called a **compound wall**. Metal ties are used to attach the two layers together and give strength to the wall. The positioning of header rows in the brick outer layer can be random, not necessarily every five to seven rows.

A masonry wall with an air space left between the inner and outer layers of the wall is called a **cavity wall**.

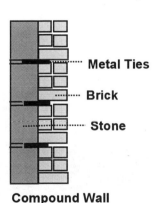

Compound Wall

In older construction, both inner and outer layers were often brick. A header row of bricks laid with the small end out, traversed the cavity and held the wall together. In more modern construction, the inner layer could be stone or concrete block. In the **compound cavity wall**, where you find different inner and outer materials, the brick could be attached to the inner layer with metal ties or with header rows.

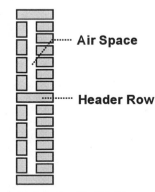

Brick Cavity Wall

Definitions

In a solid brick house, three layers or wythes of brick are used to construct a solid wall with no wood framing. Header rows are rows of bricks turned small end out to act as ties to hold the wall together.

A compound wall is a solid masonry wall built of two different materials. A cavity wall is a masonry wall with a dead air space left between layers.

Walk around the house slowly, paying attention to the entire surface of the wall and its condition at each window and door opening. Here is a list of those items you should watch for when inspecting a solid brick, stone, or concrete block wall:

Air Space

Header Row

Concrete Block

Compound Cavity Wall

Cracking in the wall indicating foundation settling and movement. Step cracks on adjacent walls at a corner is a sign of footing failure at that corner due to soil weakness. Vertical cracks that run down to the foundation often signal settlement. Jagged, sharp edged cracks are an indication that movement is active. Distortion of the window framing can also indicate settlement of the structure.

- **Cracking above openings** indicating a rotted or rusted lintel or failed arches.

- **Spalling** or **deteriorating** wall material.

- **Deterioration of mortar.** Use the tip of your screwdriver or a knife to probe mortar, testing it for condition. Note that the mortar in **Photo #7** remained intact even when the brick fell off.

- **Bowing, leaning, or distortions** in the wall caused by mortar deterioration, interior framing problems, rafter spread, and foundation movement.

#7 Spalling brick

*Take a look at **Photo #7** for an example of **spalling brick**. Here, pieces of the brick's surface are falling off due to water penetration into the brick. In some cases, a good portion of the brick has crumbled away.*

"We've come across other veneer detachment problems. Recently, one of my inspectors inspected a house where the brick veneer was in place. However, there was a considerable bow along one wall and marked displacement of the veneer around the windows and doors in that wall. He discovered that the veneer was held in place by the fascia board at the top. One push and the whole wall would have come down.

"I should say that it's better to caution customers about problems like these and not be the person who actually causes the wall to fall. It's hard to convince people that you didn't do something wrong."

Roy Newcomer

We haven't said much about **concrete block walls**. Concrete block left as the exterior wall is most often seen in commercial buildings. Although used in home construction, they're not often left uncovered. Generally, residential concrete block walls are finished with stucco (see page 18) or a layer of cement parging.

Brick Veneer

Brick houses built today, and indeed since the early 1970's, are generally brick veneer over wood framing. A brick veneer wall does not carry the load of the structure.

The brick veneer is constructed from the foundation up and is attached to the wall sheathing with **brick ties**. These ties are usually crimped, accordion-style, to allow them to expand and contract with the wooden frame and keep them from cracking the brick veneer.

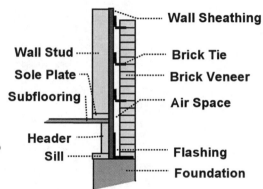

An air space of about 1" is left behind the brick veneer to allow water passing through the brick to run down the wall. This water exits through the bottom row of brick through **weep holes** (not shown here) which are openings every foot or so along the bricks. A **flashing** runs beneath the brick and up the wall to prevent this water from reaching the foundation.

With brick veneer, the inspector is able to see signs of problems. Weep holes can become blocked and not allow water to exit from behind the bricks, leading to deterioration of the brick and the mortar. Brick ties can be improperly installed or loosened over time, and the veneer can separate from the wall.

When inspecting brick veneer, the most important thing to inspect for is **detachment or separation of the veneer** from the house. The brick ties can be improperly installed or can come loose. Sight along window and door openings where you might see the lean of detached veneer — more brick shows at the top than the bottom. A detaching veneer may have a decided bow to it or can show signs of cracking as it pulls loose and separates from the house.

As with solid brick walls, the home inspector should examine the wall for **cracking** in the brick that can indicate structural problems. Check for **deterioration** of brick and mortar. Examine the **weep holes** in the lower course of bricks in the veneer to be sure they are open and functioning. Water behind the wall cannot escape if weep holes are blocked.

Often times, there is brick veneer on only a portion of the house. Be sure to check areas where the veneer meets other surfaces.

VENEER PROBLEM

Watch carefully for the separation of brick veneer from a house. Don't be the one to cause it to fall off.

#8 Example of detachment

*We wanted to share **Photo #8** with you. It wasn't one of our inspections, but it is an interesting **example of detachment**. What at first glance looks like a brick patio at the side of the house is actually brick veneer that separated from the wall and fell off. The mortar held the brick in formation. Inspecting brick veneer for detachment is something you don't want to miss. Imagine not reporting it and then getting the call from this new homeowner after the veneer fell off!*

#9 Partial brick veneer

*Photo #9 shows **partial brick veneer** on a house. Note the stone ledge at the top of the veneer. This ledge is sloped toward the house, allowing rain to pour behind the veneer. The ledge should have a positive slope. Here, you can see that the wood siding at the corner is also rotting because of this problem.*

PAINT PROBLEMS
• Chalking
• Alligatoring
• Cracks and crazing
• Peeling
• Excessive layers of paint
• Staining
• Mildew

Other Sidings

A home may have other less common wall coverings as listed here:

- **Slate shingles**: You may find slate shingles in small areas such as gables or dormers on late 1800's and early 1900's homes. Slate shingles have a very long lifetime. However, nails can rust, causing the shingles to slip out of position.

- **Fiber Cement Siding:** This has become very popular in recent years and is much more durable than the hardboard siding which has experienced many problems that resulted in lawsuits. This is a Portland cement-based siding with wood fibers. Go to your local lumber yard to view this siding.

- **Brick and stone facings:** You may find thin (less than 1/2" thick) facings on all or part of the front of the house. These types of facings are heavy and should sit on the foundation. Problems with facing include detachment and water penetration. There should be weep holes at the bottom. Otherwise, you may find moisture damage to interior walls.

A Word about Paint

The home inspector should be familiar with defects in the exterior paint surface and should know why the paint failed.

The three constituents of house paints are the **pigment** or color particles, the **vehicle** or film-forming compound, and the **thinner or solvent** which evaporates from the pigment and vehicle after the paint is applied.

The following problems may be found with a paint surface:

- **Chalking:** This is a process in which ultra violet radiation causes the vehicle to break down and pigment particles to be released. Rain washes oxidized vehicle and pigment particles away until the siding is exposed. Any chalking left on the surface should be cleaned off before repainting.

- **Alligatoring:** Wrinkling or cracking in the exterior paint surface is called alligatoring. It looks like the hide of an alligator. This condition is caused when the solvent evaporates too quickly upon application, preventing the rest of the solvent in the paint from escaping. It can happen when paint with quick evaporating solvents are applied in the sun. It can also occur if the paint is applied too thickly.

- **Cracking and crazing:** Cracks appear in the paint surface when the outer layer of paint can't respond to the expansion of the old layers of paint or the siding beneath. Cracks often appear in the paint on window sills where wet wood expands. Crazing is a network of crossing cracks all over the surface, occurring when the outer layer of paint shrinks while it is drying.

- **Peeling:** Paint peels when layers of paint and primer separate due to painting without removing chalking, painting over grease left on the surface, or new paint that won't adhere to the old. When the finish separates from the wood beneath, suspect moisture buildup in the wood. Peeling paint often indicates a leak into the siding.

- **Excessive layers of paint:** Older homes may have so many layers of paint that the layers may separate from the siding entirely and stand on their own. Moisture can get under the layers into the siding. The home inspector can check for this defect at corners.

- **Staining:** There are many causes of staining on paint, including an air pollution condition that causes a chemical reaction with the paint, rust from nails, rusting gutters and downspouts, resin or sap from the siding itself that bleeds through the paint, airborne grease and soot, and insect eggs.

- **Mildew:** Mildew is not a stain. It's live fungi that live on and in the paint, usually on the north face of the house or under shady areas where the sun can't dry it. The home inspector can test for mildew with bleach, which will destroy it but will not clean other types of staining.

Reporting Your Findings

Some home inspectors like to arrive early to give themselves time to take one walk around the house alone to get an overall view of the place, the general layout, and any obvious problems. Once the customer arrives, have the customer come with you as you start the exterior inspection. We consider having the

Definitions

Pigment in paint is the color particles. The vehicle in paint is the film-forming compound. The solvent or thinner is the third constituent in paint that evaporates after the paint is applied.

Chalking is a process in which ultra violet radiation causes the vehicle in exterior house paint to break down and pigment particles to be released.

Alligatoring is a process in which the solvent evaporates too quickly, leaving residual solvent in the paint and causing wrinkling or cracking in the surface.

Crazing of a paint surface is a condition of net-like patterns of cross cracking, caused when a new layer of paint shrinks while drying.

customer come with you as you conduct the inspection to be the number one rule in home inspection. This is for protection against lawsuits. If the customer is fully informed during the inspection, the chances are much reduced that you'll be getting a complaint call later.

It's important to be systematic about the exterior inspection. If trying to do everything in a single trip around the house with the customer causes you to miss something, plan two or three trips. It's all right to tell the customer that you'll only be looking at the foundation on the first trip, for example. You can always explain that you do this to be sure not to miss anything. The customer will appreciate it.

When inspecting the siding, keep a running dialogue going with the customer, explaining:

- **What you're inspecting** — siding, corner moldings, fastenings, joints, paint surface, weep holes, and so on.

- **What you're looking for** — deterioration, improper installation, warping, buckling, and so on.

- **What you're doing** — probing for wood rot, probing the mortar, testing for mildew, listening for loose stucco, and so on.

- **What you're finding** — chalking paint, wood rot, missing moldings, detached stucco, and so on.

- **Suggestions about dealing with the findings** — calling a mason to evaluate a mortar condition, replacing a window sill, caulking joints, repainting, and so on. But with this caution — don't make uneducated guesses about how repairs should be made.

Communication with the customer is such a major point in home inspection that it's not surprising that some inspectors fail because they can't communicate properly. Communication entails both output *and* input. The home inspector must be sensitive to the customer's attention span and their level of understanding. There's nothing worse than the home inspector who goes on and on without noticing he or she is boring the customer or who talks over the customer's head. There's no need to explain the chemistry of paint; it's enough to explain in simple terms that the paint has broken down and needs to be redone.

Filling in Your Report

Every home inspector needs an inspection report. A **written report** is the work product of the home inspection, and every home inspector is expected to deliver one to the customer after the inspection. Inspection reports vary a great deal in the industry with each home inspection company developing its own version. Some are considered to be excellent, while others are not very good at all. A workable and easy to use inspection report is important for a home inspector in terms of being able to fill it in. Of greater importance is its thoroughness, accuracy, and helpfulness to the customer. We can't tell you what type of report to use, but let's hope it's a professional one.

The **Don't Ever Miss** list presented on the next page is a reminder of those specific findings you should be sure to include in your inspection report. We list these items after years of experience performing home inspections. Missing them can result in complaint calls and lawsuits later.

When you're filling in your inspection report, be sure to put in enough detail so your customer knows what your findings were, even if the report is read at a later date.

Here is an overview of what to report on during the inspection of the siding:

- **Siding information:** Identify the type of siding present on the house, noting materials such as stone, stucco, aluminum, vinyl, and so on. You may find more than one type of siding on the house. If you do, be sure to identify each type and note its location on the house.

- **Siding condition:** In general, an inspection report should have categories of satisfactory, marginal, and poor regarding the condition of the siding. If you use the rating marginal or poor, be sure to explain further what you found troubling about the siding. It pays to put as much into the writing as you can. Here are some examples: "Stucco cracked and leaking at back of house, needs repair." "Veneer loose on east wall, needs immediate attention." "Siding should not be touching ground, need 6" clearance." "Rotted planks on north wall." "Siding needs to be repainted." "Caulk at southeast corner to prevent leaking."

DON'T EVER MISS
• Rotted wood
• Loose face brick
• Detached stucco
• Water penetration

- **Major repair or replacement:** Only in extreme circumstances will the home inspector need to classify siding as a major repair. Use this classification if siding needs to be replaced or undergo major repairs to bring it up to an acceptable level. Serious veneer detachment would qualify as a major repair. It's a good idea to report this situation on the exterior page of your report and then list the major repair or replacement again on a summary page of your report.

- **Safety hazard:** A brick veneer wall about to fall down is probably the only example of a safety hazard as far as siding is concerned. If, as in the case of our example, a simple push would bring the wall of veneer down, call it a safety hazard in your inspection report.

Report Available

The American Home Inspectors Training Institute offers both manual and computerized reports. These reports include an inspection agreement, complete reporting pages, and helpful customer information. If you're interested in purchasing the Home Inspection Report, *please contact us at 1-800-441-9411.*

WORKSHEET

Test yourself on the following questions.
Answers appear on page 30.

1. Identify the type of siding shown in Photo #5 at the back of this guide.

   A. Clay shingles
   B. Asphalt composition
   C. Channel planking
   D. Composition board

2. What is <u>not</u> the purpose of siding?

   A. To enhance the rigidity of the framework
   B. To protect the framework from the elements
   C. To protect the interior of the structure from the elements
   D. To carry the weight of the roof and floors down to the foundation

3. A permeable wall is one which:

   A. Allows moisture to pass through it.
   B. Has insulation between the studs.
   C. Is air tight.
   D. Has polystyrene foam board insulation.

4. Which pair of sidings would the home inspector most easily recognize by their <u>differences</u>?

   A. Composition board planks and clapboard
   B. Vertical butted planks and shiplap planks
   C. Vinyl siding and asbestos cement siding
   D. Painted clay shingles and painted cedar shingles

5. Double coursing means that:

   A. Two layers of shingles are laid for each row.
   B. Alternating courses of shingles are laid at different exposures.
   C. More than 2" of shingle is exposed to the weather
   D. Half the shingle is exposed to the weather.

6. What is a cause of shingles buckling?

   A. The joints too tightly spaced
   B. Joints not staggered from row to row
   C. Butt edge not flat cut
   D. Loose nailing strip

7. What is the number one cause of problems with aluminum siding?

   A. Siding not grounded
   B. Holes at bottom of planks plugged
   C. Defective installation procedures
   D. Lack of sheathing underneath

8. How should the corner moldings on aluminum siding be installed?

   A. Each molding butting the one above and below
   B. The upper molding overlapping the one below
   C. The lower molding overlapping the one above
   D. Any of the above

9. Which statement is <u>false</u>?

   A. Stucco applied to concrete block does not need a layer of wire mesh.
   B. Stucco is water resistant and impermeable.
   C. Trapped moisture can cause stucco to detach from the wall.
   D. Random cracking of stucco can occur during curing.

10. What is the process in which the vehicle in paint breaks down and releases pigment particles?

    A. Crazing
    B. Alligatoring
    C. Chalking

Pages 30 to 38 present the study and inspection of exterior trim, windows, and doors.

Definitions

The cornice consists of the trim and moldings at the eave line. At the gable end of the roof, it may have only a vertical board called the rake-board or barge rafter. A closed cornice has both a vertical fascia board and a horizontal soffit on the underside of the eave.

Chapter Three

TRIM, WINDOWS, DOORS

Another aspect of the exterior inspection is to inspect the trim, the windows, and the exterior doors of the house. The home inspector may include the inspection of these items while inspecting the siding or make a separate trip around the house.

Inspecting the Trim

The exterior trim includes all pieces added to the siding that serve to protect the framework and the interior of the structure from the elements. The trim seals and protects the joints where the siding ends. Trim includes casings around windows and doors, corner boards, skirt boards at the bottom of the siding, the **fascias** and **soffits** that make up the side and under part of the eaves. The eave trim is also called the **cornice**. Shutters and window boxes should be examined for their condition and for their possible damaging effect on the siding behind them.

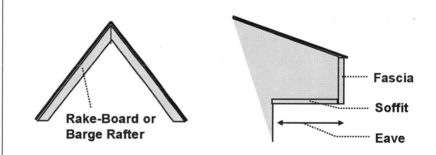

- **Wooden trim** is likely to be pine which needs regular painting and maintenance. Look for the following:

- **Wood rot:** The home inspector can use a probing tool attached to a long pole for reaching and probing any first floor fascias and soffits from the ground. Get up on the ladder and use the pole to reach higher areas. Test probe a representative number of areas and probe all suspicious areas where wood rot is suspected.

If there is rot found in the trim, examine the area beneath where the siding, sheathing, and framing may also be rotting. Rot can spread when rotted trim is merely painted over. It's difficult to discover that condition without probing in random areas.

Worksheet Answers (page 29)

1.	B
2.	D
3.	A
4.	C
5.	A
6.	A
7.	C
8.	B
9.	B
10.	C

- **Condition:** Look for warping, split, and broken trim pieces. Watch for insect damage. Soffits and fascias can be damaged by small animals such as squirrels and birds trying to get into the attic.

- **Loose or missing trim:** Check to see that all trim pieces are present and securely attached. If missing or loose, there is probably water leakage into the wall. Check seams and joints in the trim pieces. Watch for nails coming loose.

- **Painting and caulking:** The condition of surface paint should be inspected (see pages 24 and 25). Let the customer know that they should expect to maintain trim on a regular, perhaps annual, basis. Be sure to point out areas that need repainting as a part of a regular maintenance routine, distinguishing them from peeling areas that indicate serious water penetration into the trim.

  Check caulking at seams and joints in the trim and at interfaces around windows and doors. Inform the customer that caulking is also part of routine maintenance and should be examined on a regular basis.

Aluminum and **vinyl trim** are, for the most part, maintenance free. But the home inspector should examine it to make sure that the pieces are properly installed, no pieces are broken or missing, and the trim is still securely fastened.

If exterior **wooden shutters** are present, examine them and tell the customer about the presence of wood rot or other deterioration. Try to assess what is going on with the siding behind deteriorating shutters. Sometimes the condition of the shutters can have a damaging effect on the siding they cover. When looking behind shutters, be careful — wasps like to build nest in places like this.

Check window boxes for condition and examine their hangers and supports for decay.

There may be **miscellaneous structures** attached to the outer wall such as decorative boxes holding lighting fixtures and niches in the wall. Examine each for deterioration. Boxes holding any electrical work must be waterproof, so check weatherproofing and make sure the box is firmly attached to the wall. Check caulking around such structures.

INSPECTING TRIM

- Wood rot
- Broken or split pieces
- Loose or missing pieces
- Loose nails
- Painting and caulking
- Proper installation

Definitions

A window sash is the frame-work that holds the glass or other material. The window frame, which surrounds and holds the sash, is made up of a top piece called the head, side pieces or jambs, and a bottom piece or sill. The window casing covers the edge of the window frame where it meets the wall covering.

A J channel is a manufactured component of an aluminum or vinyl siding system which has a curved channel that the planks fit into. J channels are used around window and door openings to make a weathertight seal.

Window Construction

All exterior window components are inspected during the exterior inspection. Although most standards of practice state that a representative number of windows must be checked for *operation*, that's left to be done once the home inspector actually goes inside the house.

The components showing on the exterior of the window are the **sashes** (upper and lower in the traditional window), the **head** and **jambs**, the **sill** and the **subsill**, and the **casings** around the window.

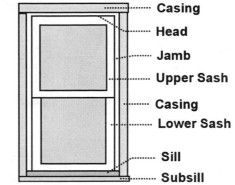

For a weatherproof opening in the exterior wall, building paper and/or adhesive flashing is laid around the window. Ideally, flashings are present at the top of the window (see drawing on page 7). Vinyl and aluminum siding may be installed with **J channels** around openings that the siding fits into, forming a weatherproof seal. If not, caulking is required around windows the same as it is with other siding materials.

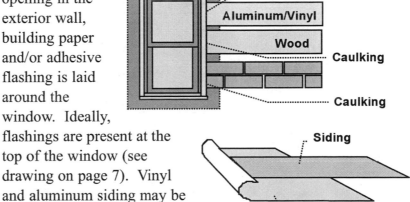

There are many styles of windows. The most common is the **double hung window** with two sashes that move. The upper sash is on the outside; the lower on the inside. Windows may be **single hung** where only the lower sash moves. A **slider window** is one with a sash that moves horizontally.

Sashes can be hinged into window framing to open in a variety of ways. The **awning window** is hinged at the top and opens outward. A **hopper window**, often found in basements, is hinged at the bottom and opens inward.

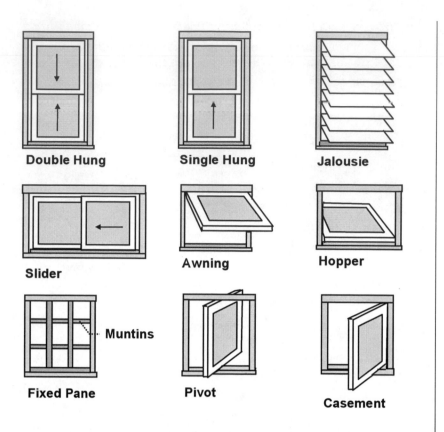

Double Hung Single Hung Jalousie

Slider Awning Hopper

Muntins

Fixed Pane Pivot Casement

SLIDING WINDOWS

- Single hung
- Double hung
- Horizontal sliders

HINGED WINDOWS

- Awning
- Casement
- Hopper
- Jalousie
- Pivot

MULTIPLE WINDOWS

- Combination
- Bays and box bays
- Bow windows

The **casement window** is hinged at the side to open outward. The **pivot window** pivots from a center hinge. The **jalousie window** contains narrow strips of glass in a device that allows the strips to move together, lifting out from the bottom.

A **fixed-pane window** is one that does not open or close. A **picture window** is basically a large fixed-pane window. **Combination windows** can be made up of a large fixed-pane window in the center between two smaller casement windows. Another configuration might be a fixed-pane window in the center with a slider at each end. Fixed-pane windows may have snap-in **muntin bars** (a grid of crossed pieces of wood or plastic creating a simulated divided light window) that fit into the window. A true **divided light window** has small individual pieces of glass set into wood or lead muntins.

Bay windows are made up of three windows set at angles with each other in a bay that protrudes from the structure. There is usually a larger fixed-pane window on the length of the bay with standard size, opening windows at each side. In a **box bay**, the windows are at right angles. A **bow window** is similar to a bay, but has more than three windows, each at angles from the others.

Definition

Muntins are a grid of cross pieces of wood, plastic, or lead that hold small panes of glass in a divided light window. Snap-in muntins made of wood or plastic can be applied to imitate a true divided light window.

Glazing is a term that refers to the window pane made of glass or other material. Each layer of glass is called a **light**. Therefore, the glazing may be made up of a single light or single pane of glass or of **multiple panes**, where two or three thin pieces of glass are sealed together.

Insulated glass has a space between the two lights which is often filled with an inert gas such as Argon. Triple lights may be three layers of glass or two outer layers of glass with a plastic layer in-between. Sometimes an inner light has a reflective Low –E coating (E for emissivity) meant to control heat gain and loss.

When multiple pane windows are cracked or the seal is broken between layers of glass, air and moisture leak into the layers. Condensation can be seen between the layers. Leaking multi-pane windows will discolor and should be replaced.

Glazing can be **transparent** (clear) or **translucent** (opaque or cloudy) by design. The home inspector may come across plastic, acrylic, and polycarbonate glazing as well as glass.

The home inspector is *not* required to report on the effectiveness of **safety glazing**. However, for your information, safety features include the following:

- **Tempered glass:** This type of glass is difficult to break. When it does break, the pane shatters into small, smooth edged cubes.

- **Laminated glass:** A sticky plastic inner layer between panes holds broken pieces when the pane cracks or shatters.

- **Wire mesh:** Some windows had a wire mesh embedded between the panes for security reasons, not so much for protection from broken glass.

Inspecting Windows

When inspecting windows, remember that the window is an interruption in the wall structure and an opening in the siding system. The interruption to the structure must be supported by a header between the studs. As an opening to the siding, the opening must be properly sealed and weatherproofed.

The home inspector should pay attention to the following aspects of windows during the exterior inspection:

- **Header:** Watch for sagging headers above large windows. Rusted and rotting lintels can cause cracking above the window and allow water to leak into the window casings and frame. Probe wood at the top of windows for wood rot.

- **Flashings:** Check for the presence of flashings above windows. Flashing is not always there. Instead, there may be a wood drip cap molding over the top of the window casing. These moldings can rot easily. They're often unpainted on the back, and water, instead of being deflected off the edge of the molding, seeps behind the back of the molding. Probe them for wood rot. Windows under overhangs should be protected from these problems.

- **Framing:** Inspect wood casings, frames, and sashes for wood rot. The sill is especially vulnerable to rot and should be probed even if a new layer of paint covers it completely. Note other damage such as loose or detached framing pieces, loose nails, deteriorating paint, rusting on metal framing, and insect damage. You may want to refer back to **Photo #3** for an example of a rotted sill.

- **Glazing:** Inspect glass and report broken, cracked, and discolored window panes that may indicate a broken seal. Check that the putty or glazing compound holding the glass into the sash is not loose or missing.

- **Caulking:** Examine caulking around the casings to make sure the window is sealed where it meets the siding. Caulking should not be loose, missing, cracked or broken. When J channels are present, check to see that they form a seal with the window opening.

- **Level and plumb:** If window frames are distorted and no longer form right angles at the corners, there may be defects in the structure. Windows pulled out of square can be the result of more serious structural problems such as foundation settlement. Try to find the cause of any distortion you find.

Personal Note

"For the house with wood shingles in Photo #3, I got up on the ladder to check out that window. The casings and frames were rotted out too.
"I realized I would have to inspect every window on the house. Some didn't look too bad, but once I got up there I could poke my finger at them, and it would go right through. There were a total of six windows with rotting framework and casings."

Roy Newcomer

Personal Note

"Remember the construction worker I mentioned who brought home different siding materials for building his own additions? Well, this same fellow got his windows from the job too. The inspector had to inspect each one, not just a representative number of them, because every single window in the house was different."

Roy Newcomer

- **Storms and screens:** Where weather is cold, storm windows are used to protect the home against heat loss. With many types of windows such as double hung, storms are attached to the outside of the window. Hinged windows opening outward have storms for the inside. Examples are casement, pivot, and jalousie windows.

The inspection of storm windows, although not required by most standards of practice, should describe both the materials used in construction and their condition. Report storms with cracked or broken glazing and deterioration of materials or putty. If storms are present on the house, make a note if any windows are missing them.

If condensation appears between storm and primary window, the home inspector can tell which of the two windows is leaking. If moisture appears on the storm, then the inner window is leaking air from the house. If moisture appears on the inner window, then the storm is leaking air from the outside.

If screens are present, check them quickly for general condition, paying attention to those that are bent or torn.

Inspecting Doors

All outside doors should be inspected and *operated* during the home inspection. Entry doors can be single hinged doors, French doors, or sliders such as a patio door and window combination. Some doors are solid wood, vinyl, fiberglass, or steel, while others have partial windows or have large lights set in a frame. The entryway may have side lights (small windows on one or both sides of the door) and a transom. There may be only a single door or the addition of a storm door.

NOTE: Any entry door should be fire resistant. Some area codes require a fire rating on these doors. There may be a manufacturer's label on one edge of the door indicating the rating.

The inspection of exterior entry doors includes the following:

- **Condition:** Pay attention to the door itself, noting whether it's solid, paneled, made of wood or metal, and is partially or completely glazed. Make note of denting in the surface of the door, warping, cracked or broken panels, and deteriorating finishes. When glazing is present, inspect for broken glass and discoloration indicating a leak in the seal. Check the condition of the putty or glazing compound.

- **Operation:** Open each entry door and check its operation. Inspect hinges on **hinged doors**. They should be tight and large enough to support the weight of the door. If the door rubs the frame, loose hinge screws can be the cause. Swing the door open and watch for easy movement. If weather-stripping is present, inspect it to be sure it is firmly attached. Check the lock hardware to be sure it works.

  With **sliders**, slide the door open and check that the bearings are running on the metal tracks smoothly. Sliders that stop halfway along the track may have dirty or worn tracks and bearings that don't work any longer. Movement can be impeded by distortions of the frame and by a sagging lintel bearing down on the door. Check that the lock hardware is working.

- **Framing:** Inspect casings and door framing for wood rot. A door has, in addition to its head and jamb framing, additional framing pieces called **stop moldings** that restrict the movement of the door. At the foot of the door is the **threshold** that keeps out the rain. Around this additional framing is the casing. All of these pieces should be inspected for wood rot and for broken, loose, or missing pieces. The threshold acts like the window sill and is particularly susceptible to wood rot.

- **Caulking:** Doors should be caulked the same as windows. Inspect the caulking and make sure it's secure, not broken, and not missing in any areas. Check J channels too.

- **Header:** Watch for sagging headers, especially double doors and wide openings such as sliding door and window combinations. Rusted, rotted, and broken lintels can cause the doorframe to distort and door operation to be impeded.

- **Level and plumb:** As with windows, distortions of the door frame can indicate structural problems with the house. Use a level above large expanses of doors and windows to check that the line is level.

INSPECTING DOORS

- Condition
- Door operation
- Framing
- Caulking
- Header
- Level and plumb
- Storm and screen doors

ABOUT SECURITY

It's best that the home inspector refrain from commenting on how secure a house is. Report malfunctioning locks and door hardware, but don't comment on how good you think the locks are. Security and level of risk is a personal attitude.

Personal Note

"Sometimes the little things you notice make a good impression on the customer. Pointing out a torn screen, trying the doorbell for operation, and testing door locks convinces them that you're paying attention to every little detail. That's why we suggest some items beyond the normal standards of practice. Let's say it's because of salesmanship.

"It's not that you have to write up these details, but that they become part of your ongoing conversation with the customer."

Roy Newcomer

• **Storm and screen doors:** Depending on the season a storm door or screen door or combination storm/screen door may be present. Storms and screen doors are installed on the outside of the stop moldings and open outward. They can be made of metal or wood. Although not required, we advise the inspector to examine storm doors and screens and make note of any defects. Slide open outer screen doors that accompany sliders and point out to the customer any faulty operation. Report if screening is torn or loose.

Reporting Your Findings

Sometimes it seems as if you can't possibly write everything in the inspection report except the important findings. Yet you'll be seeing many small details when you're inspecting the windows and doors that should be mentioned. These minor points should at least be *talked about* with the customer as the inspection is being conducted. You see how important communication is.

As you record your findings of the trim, window, and door inspection, be sure to record the following:

• **Trim:** Identify the materials used in trim, soffits, and fascia boards. Then report on their condition, noting defects such as wood rot.

• **Caulking:** Your report should have room to record your findings on the condition of the caulking. Identify its general condition as satisfactory, marginal, or poor.

• **Windows:** Identify the type of windows present and the materials used in the window framing. Rate the general condition of the windows and be sure to record your comments on specific defects you've found such as rotted sills, cracked glazing, and so on. Also, we suggest that you identify the type of storms and screens and their condition.

• **Doors:** Take the trouble to list each exterior door (front entry, patio door, and so on) and write your findings for each one. For example, the front entry door may be in perfect condition, but the back door may have a serious defect. A general rating applied to all doors is not detailed enough.

WORKSHEET

Test yourself on the following questions.
Answers appear on page 40.

1. Identify the type of hinged window in each of those shown below.

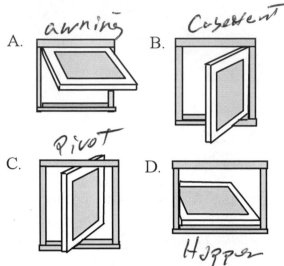

A. *awning*
B. *Casement*
C. *Pivot*
D. *Hopper*

2. What is the soffit?

   A. The trim piece that appears along the top of a wall at the gable end of the roof
   B. The underside of the eave
   C. The board laid vertically at the outer edge of the eave
   D. The portion of the roof that extends beyond the outer wall

3. What is the window sash?

   A. The outermost window framing that meets the siding
   B. The glass within the window
   C. The framework that holds the glass
   D. The head, jambs, and sill

4. Which statement is <u>true</u>?

   A. J channels are used with some aluminum and vinyl siding systems.
   B. J channels should be caulked.
   C. J channels are used with brick siding.
   D. The absence of J channels with wood siding should be reported.

5. Discoloration in a large picture window is an indication of:

   A. Translucent lights
   B. Laminated glass
   C. Poor installation
   D. A leaking seal

6. Which statement is <u>false</u>?

   A. A home inspector should give the house a security rating.
   B. A home inspector should inspect the lock hardware.
   C. A home inspector should not pass judgment on how secure the house is.
   D. A home inspector should determine if hinge screws are securely fastened.

7. A hinged door that rubs the frame during operation may have:

   A. A missing stop molding
   B. Loose caulking
   C. Loose hinge screws
   D. Missing weather-stripping

8. What areas should be probed for wood rot?

   A. A representative number of areas along the eaves
   B. Trim, window, and door pieces with deteriorating paint
   C. Window sills with a new paint job
   D. All of the above

9. Identify the type of window shown here.

   A. Double hung
   B. Box bay
   C. Pivot
   D. Jalousie

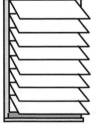

*Pages 40 to 55 present
procedures for the inspection of
the many other exterior
elements that must be inspected.*

Definitions

*The_riser is the vertical portion
of a step. The_tread is the
horizontal portion of the step.*

*The stringer is the side
supporting member that
supports the stairway.*

*Balusters are the vertical poles
that support the railing of a
staircase.*

Worksheet Answers *(page 39)*
1. *A is an awning window.*
   B is a casement window.
   C is a pivot window.
   D is a hopper window.
2. *B*
3. *C*
4. *A*
5. *D*
6. *A*
7. *C*
8. *D*
9. *D*

Chapter Four

AROUND THE HOUSE

In addition to the siding, trim, windows, and doors, many other items are part of the exterior inspection of the house. The following other exterior items will be studied in this chapter:

- Attached structures such as steps, balconies and decks
- Driveways, walks, and patios
- Grading that affects the foundation
- Retaining walls

Steps and Stoops

The home inspector should inspect *all* exterior stairs and stoops to decks and entryways. Codes vary across the country regarding handrail requirements, dimensions of risers and treads, and other regulations. The home inspector would be well advised to check with local building codes for information.

Risers are the vertical portion of the steps. In general, risers should be all the same height. Heights allowed are generally from 7" to 8". Some steps are constructed with open risers, that is, with the vertical below each step left open. Codes in some areas forbid open risers. The **treads** are the horizontal portions of the steps. They should be of even depth, generally from 9" to 11". The treads are allowed to extend over the riser 1 1/4". Codes also vary about **handrails**. Generally, when there are more than three steps or the height is greater than 24" to 30" from the ground, a handrail is required. Where steps are wide, a handrail may be required on each side.

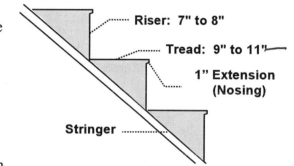

Riser: 7" to 8"
Tread: 9" to 11"
1" Extension (Nosing)
Stringer

When inspecting exteriors steps, the home inspector should inspect the following items:

- **Platform at the top:** When steps lead to an entry door, the platform at the top should be large enough to stand on while opening an outward swinging storm door or screen door. If the platform is too small, this should be reported as a **safety hazard**.

- **Condition of materials:** With **wooden steps**, the first requirement is that the steps are sturdy enough to support the weight of a person on them. Steps should not be too flexible or springy. Inspect for rot, especially at the feet of the stringers and any area that may be in contact with the soil. The home inspector may advise the customer to remove outdoor carpeting from wooden steps because it holds moisture and promotes decay. Make sure all fastenings are tight. Report any dangerous conditions such as rotting boards as a **safety hazard**.

  Steps may be made of **stone or brick**. Inspect for loose stones and bricks, and check to see the mortar is in good condition. Loose stones or bricks should be reported as a **safety hazard**.

- **Risers and treads:** Inspect risers and treads for appropriate dimensions and pay attention to whether risers are of uniform height and treads of uniform depth. Unexpected differences from one step to another can pose a trip hazard and should be reported as a **safety hazard**.

- **Handrails:** Inspect handrails carefully. Know your local code as to when handrails are required. Handrails should be of the proper height — roughly 32" to 38" high, but varying according to local codes. **Balusters**, which are the vertical poles in the railing, should be no more than 4" apart to prevent children from getting through. Check that handrails are securely fastened at the bottom and where they connect to the house. Missing and/or loose railings and balusters should be reported as a **safety hazard**.

Stoops can be made from brick, stone, concrete block, or poured concrete. Precast concrete stoops are available, brought to the site and mounted on concrete piers that extend below the frost line. The home inspector should inspect stoops for cracking which can present a trip hazard. Check to see if mortar is in good condition.

The soil below a stoop can settle, causing the stoop to settle or tilt along with it. If the stoop tilts toward the house, it can press against and damage the foundation. A stoop tilting significantly away from the house can be dangerous to walk on. Customers should be advised that either of these situations should be fixed.

SAFETY HAZARDS

- Platform too small
- Rotting boards
- Loose stone or brick
- Uneven risers and treads
- Loose or missing handrails
- Balusters too far apart
- Cracking or tilting stoops

For Your Information

Find out what your local code is regarding exterior steps and stoops.

Photo #10 shows a *concrete block stoop* which at first glance appears to be settling and tilting. But notice that this stoop is mounted on piers at its far edge. The stoop is level and remains square to the house and has not settled. What's settled are the fill blocks at the sides. Notice, however, that as the fill blocks move down, the supporting block above is beginning to break up, especially right next to the house. There's a danger here that the block next to the foundation will come loose and the stoop will loose its support at that end. We pointed that out to the customer as a problem in progress.

#10 Concrete block stoop

Photo #11 shows a *combination stone and wood porch* with an entry panel at the left side. Although it's not a great experience to crawl into spaces like these, we entered through that panel and inspected the underside of the porch. To our surprise the space under the porch extended across the front of the house beneath the bay at the right. The room had been built outside the foundation and was really an enclosed porch, explaining why the room was almost impossible to heat.

#11 Combination stone and wood porch

Porches

A porch is a roofed extension of the house and built as part of the house, as opposed to a deck which is a separate attached structure. When inspecting a porch, inspect the porch stairs for the conditions and hazards as outlined on the preceding pages. Also inspect for the following:

- **Supports:** Porch **columns** and **posts** support the roof above and the floor system. They may be wood, metal, poured concrete, brick, or stone. Columns and posts should have their own footings to prevent settling. The home inspector should look under the porch to determine if the support materials are in good condition, inspecting for wood rot, mortar deterioration, settling, and so on. The **floor joists** and **beams**, should be inspected from underneath to determine whether they're properly spanned to support the dead and live loads on the porch floor, in good condition, and properly supported at their ends.

- **Roof structure:** The home inspector should eye the roof of a porch to spot sagging. The ceiling of the porch will most likely be covered with finishes such as plastering, stucco, drywall, or wood. But the inspector should look for sagging in the ceiling that indicates leaking into the roof structure or beams that have warped or lost their bearing on the support columns. Be sure to probe ceilings and fascias for wood rot.

- **Floors:** Open porches are built with a floor that tilts away from the house to allow rain to drain away. Porches that are later enclosed often keep the same pitched floor. Examine floors for wood rot. Just as with wooden steps, a wooden porch floor should not be covered with outdoor carpeting as it holds moisture and promotes decay.

- **Railings:** Know your local code as to when porch railings are required — generally about 24" to 30" or more from the ground. Baluster requirements vary also, but their spacing should be 4" or less. Be sure to report any variation from these requirements as **safety hazards**. This may seem like a minor detail. However, toddlers can squeeze their entire bodies through openings of less than 5", and there are tragic stories about children getting their bodies through balusters and actually hanging themselves. So be serious about reporting safety hazards.

INSPECTING PORCHES

- Deteriorating supports
- Sagging and leaking in roof and ceiling
- Decaying floors
- Safety hazards in steps and railings

Decks

A deck is basically a platform that is attached to a house, but is an independent structure all by itself. Decks can be completely open to the weather, as is most common, or roofed, partially or completely walled or screened. The key to identifying if such a structure is part of the house structure or independent is in the supports. A deck is not built on the house foundation; it is usually mounted on columns and posts.

With decks, inspect the following items as described for porches in the preceding text:

• Steps
• Supports
• Floors
• Railings

A special aspect the home inspector should pay attention to with **deck supports** is that wooden support posts should have their feet above the ground. Probe the wood below the ground for wood rot and insect damage if you find them buried. If possible, check beneath the deck for the presence of wood rot in the joists. Some decks are built so close to the ground or are entirely closed in that ventilation is poor and decay is promoted.

The wood used for decking is often pressure treated, protecting the wood from termites and rot, but staining and the use of water repellents are suggested to make it truly weather resistant. Point out decks that need re-staining and water repellent treatments.

Floor boards not properly maintained can cup and check, holding more water as the process progresses. There should also be space left between the floor boards when laid to prevent water from being unable to evaporate at the seams. Check carefully for wood rot in flooring and also at the header joist at the perimeter of the deck.

The **fastenings** by which the deck is fastened to the house should be inspected. Fastenings should be secure so the deck doesn't separate from the house. There may be nails, metal hangers, bolts, or lag screws (lag bolts should be recommended if not present). Inspect the area for wood rot.

#12 New dock sitting at a building center

Photo #12 shows a **new dock** sitting at a building center. Notice anything wrong with it? Although the balusters are properly spaced, there is a space larger than 2" below the balusters. Docks are not typically part of a home inspection.

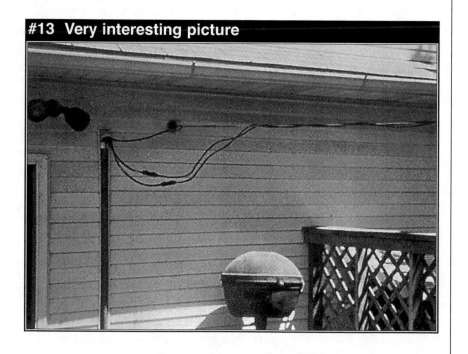

#13 Very interesting picture

Photo #13 is a very interesting picture. This homeowner had added a new deck and a door from the house to the deck. But he didn't have the power company change the position of the electrical work on the house. Now the wires are at a very dangerous height. Just imagine someone cooking at the grill, losing their balance, and grabbing the wires for support. The point is to keep your eyes open when examining decks. Because they're so often added later, all sorts of mistakes can be made.

INSPECTING BALCONIES

- Safety hazards on railings and balusters

- Water penetration at junction with house

- Wood rot

- Shaking, sagging, and tilting

Balconies

A balcony is defined as a platform protruding from the house that may not be supported by the ground. They can be built on cantilevered joists extending to the exterior or on joists fastened to the interior framing. Some balconies have exterior supports such as cables or wood that essentially supports the balcony from above. All framing members on balconies should be inspected for **wood rot** — supports, joists, flooring, railings, and balusters.

Problems are common where the balcony meets the wall of the house. Beams and joists cantilevered out can crush from the weight of the balcony at this point. Water often penetrates into the intersection and promotes decay within the wall.

Be cautious when stepping out onto a balcony. They're often not used and can be neglected. Test it first for **stability**. In general, if you notice shaking, sagging, or tilting of the balcony when stepping out onto it, report the condition as a safety hazard.

*Photo #14 shows a **balcony**. What's wrong here? Right, the railing and balusters (if you can call them that) certainly don't meet code. This situation is an obvious safety hazard. This cantilevered balcony also showed evidence of leakage into the interior framing; interior walls and ceilings should be examined. Note that some battens on the siding aren't stained.*

#14 Balcony

Patios

A patio is a flat, paved area abutting the house, made of stones or bricks or a poured concrete slab. A patio should **slope slightly away** from the house to prevent water from running toward the foundation. Inspect stone and brick patios for **trip hazards** —pieces tilting up with a raised edge, pieces missing, surfaces deteriorating. In concrete patios, there shouldn't be any unplanned cracks. Offset cracks in the surface should be reported as a trip hazard. If patios have railings, fencing, or side walls or brick or stone, inspect them for condition.

#15 Concrete patio

*Photo #15 shows a **concrete patio** that is pitched toward the house. The row of bricks placed along the house was the owner's attempt to stop what was happening here. It didn't help, of course. Water had been flowing toward the house and into the basement. Notice the debris washed against the house.*

Walks and Driveways

Walkways can be made of concrete, bricks, pavers, gravel, or flagstones. Sidewalks on the property should be inspected for their condition and whether they pose a **trip hazard**. Always report trip hazards. The sidewalks can have cracking or uneven settlement that can easily trip people.

Sidewalks near the house should not slope toward the house, especially those that abut the foundation, for obvious reasons. Water should flow away from the foundation.

Photo #16 shows a *concrete sidewalk*. Notice the three slabs that have settled and how they're offset from the slabs on either end. That's a trip hazard. And notice the pitch toward the house. In this case, the slope around the house is such that water from the sloping sidewalk can't reach it. Water wasn't reaching the basement. But water flowing from the house at the left and water flowing across the sidewalk from right to left pools in the depression. The trip hazard, settlement, and slope toward the house were pointed out to the customer.

#16 Concrete sidewalk

INSPECTING WALKS AND DRIVEWAYS
• Trip hazards
• Negative slope
• Pitting, upheaval, and sunken areas

Driveways may be dirt or gravel or be paved with brick, pavers, asphalt or concrete. Driveways should be inspected and the following types of problems noted:

- **Pitting** can occur as a result of poor quality of materials, poor installation, acid spills, or the use of salt as a de-icer. Extensive salting in cold climates can disintegrate an old or inferior concrete surface.

- **Upheaval** can be caused from poor construction, an insufficient base, tree roots, and even stones that move to the surface after years of frost and heaving. If the condition is extensive, a complete replacement may be called for, especially if a construction deficiency is the cause. If tree roots are causing the upheaval, removing the tree will eventually halt the process.

- **Sunken areas with cracks and breaks** are caused by uneven soil settlement. If the damage is not too severe, mudjacking the paved area may solve the problem. The backfill material should be clay soil or gravel.

Landscaping and Grading

During the exterior inspection of the house, the home inspector should pay attention to the **vegetation** and the **grading**, or slope of the land, around the house. The basic concern during the home inspection is to determine whether landscaping and grading cause damage to the exterior and the foundation. The principles are simple:

- **Principle #1:** Vegetation and grading should not encourage water to flow toward the house.

- **Principle #2:** Vegetation should not be allowed to damage siding, trim, and roofing or pose the potential of doing so.

As you inspect the exterior of the house, take note of the slope of the land at the foundation. The land should slope away from the house. A proper slope is 1" per foot over 5' or 6' from the house. Land with a reverse (negative) slope sends excess water toward the foundation and eventually ends up in the basement.

In some cases, adding additional backfill to slope the land away from the house solves the problem. That may pose an additional problem at the basement windows, which would then be below grade. **Window wells** may be recommended to prevent water penetration through the windows.

#17 Reverse grading

Photo #17 shows *reverse grading* toward the foundation. A situation like this must be brought to the attention of the customer and recorded in your inspection report. The condition is not easy to fix. Regrading would require so much soil to be removed that the grade would then lie well below the level of the driveway and entrance. The home inspector can recommend that a soil engineer or landscape architect be called in for a consultation. One suggestion for solving the problem was to keep the slope going towards the house as is, then trench out an additional slope along the house (toward the camera). Water running toward the house could be made to run down along the perimeter and then away from the house.

"We were asked to inspect the house in Photos #18 and #19 when the owners brought a lawsuit against the builder. I don't think I've ever felt quite so sorry for any young couple as much as this one. The builder had messed up so badly and resisted the only workable solution because of the expense. He had suggested digging out the land around the house, which would have had the result of this house sitting in a hole with an elevated driveway and garage. What a jerk."

Roy Newcomer

If window wells are already in place, they should be inspected for drainage, for the presence of a good gravel base, for corrosion of the metal well siding, and for absence of debris buildup. If there's a water line on the basement window, you'll know that drainage is poor . Some window wells have plastic domes to deflect rain away. Inspect the domes for cracking and breakage. If metal grills appear over the window well, inspect them for corrosion and breakage. Report grills with sharp cutting points and those that are corroded as safety hazards.

Photos #18 and #19 show a very serious **grading problem**. Take the time to study these pictures carefully and see if you can determine what's going on.

#18 Grading problem

First look at **Photo #18**. The slope from the house to the driveway appears to be pitched correctly. However, the bottom board of the siding is buried, and it shouldn't be.

#19 Another Grading problem

*Now look at **Photo #19**. Where the soil is even with the driveway the bottom board is buried. Note where they've tried to slope the soil off to the right,this is a major problem. What had happened was that the builder dug the hole too deep and then didn't compensate by adding another row of concrete block to the top of the foundation. The top of the foundation is actually below grade. The builder tried to cover the problem by laying the driveway and garage slabs higher than the foundation and backfilling above the foundation. The only possible solution to the problem is to raise the house and add that extra row of blocks.*

Make one trip around the house just to look at grading and vegetation. Note the presence of **trees** close to the house. Trees too close can lead to root problems with the foundation and sewer lines, messy gutters, and falling overhanging branches. Where areas are too shady, siding may not have the chance to dry out and moss and mildew can thrive. Point this out to the customer and suggest that trees be trimmed back or removed if they present a danger.

Vines on the house may be cozy, but it's not a good idea. They can hold moisture and promote insect damage. English ivy has a very strong grip and can puncture paint surfaces, grow behind siding and loosen it, and even grow under sills. Vines keep the siding from drying out. The home inspector should inspect the siding for damage. If there is none, it should be recommended that the customer monitor the situation.

Personal Note

"Keep your eyes open when walking around the house looking at vegetation. One inspector I'm acquainted with stepped into a pile of leaves only to discover that the leaves were hiding a tarp lying over a 6' deep hole. He broke his leg as a result."

Roy Newcomer

Shrubbery near the house should be trimmed back so there is about a foot clearance from the house to prevent moisture retention. Loose and mulched soil in **flower beds** should not touch wood siding or cover the top of the foundation. Leaves and plant debris should be raked away from the house.

#20 Stone retaining wall

*Photo #20 shows a **stone retaining wall** that should be inspected. It's close to the house and is holding soil in place against the foundation. If this wall gives way, soil will follow and open up holes where water can get at the foundation.*

Retaining Walls

The home inspector should inspect those retaining walls on the property that **affect the house and/or garage.** Other retaining walls do *not* have to be inspected.

- **Wooden retaining walls:** Horizontal walls of wood are fairly common in residential construction. The walls are usually built to lean back toward the high side. Wood members are connected to each with metal spikes. Gravel fill is added behind the wall; the soil above meets the wall to allow water to run over.

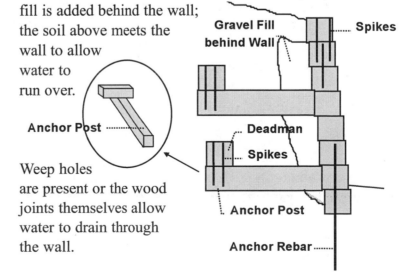

Gravel Fill behind Wall

Spikes

Anchor Post

Deadman

Spikes

Anchor Post

Anchor Rebar

Weep holes are present or the wood joints themselves allow water to drain through the wall.

Definitions

A tie-back is used to anchor a wooden retaining wall to the soil behind it. Each tie-back has an anchor post which connects to the wall and extends back into the soil. A deadman is a cross piece spiked to the anchor post at the end in the soil.

The wall is anchored to the soil through the use of **tie-backs** with an **anchor post** and a **deadman** cross piece. The tie-backs occur staggered between every second or third vertical pier. Vertical anchors are used with walls over 30" high. These anchors are placed about every 10' along the wall and extend about 4' into the soil.

- **Precast concrete:** There are new wall systems on the market made of interlocking concrete sections that also make use of tie-backs. These blocks come with various sizes and shapes with decorative stone-like surfaces.

- **Poured concrete:** A retaining wall made of poured concrete reinforced with steel can be like an inverted T, where the cross piece is buried below the soil to prevent the wall from tilting forward. Such a wall would have its footing below the frost line to prevent heaving and be equipped with drain holes near the bottom.

- **Masonry walls:** The home inspector will find masonry retaining walls too. Often times, these types of walls, especially those of stone, are dry laid and move out of position rather easily.

- **Pile walls:** Normally, vertical steel members are driven into the ground until able to resist the pressure of the soil beneath them. Wood pile walls can work in sandy soil or gravel.

Water is the main cause of **retaining wall problems**. When inspecting retaining walls, first look behind the wall for holes or depressions in the soil that can pool water. The soil should come to the top of the wall to allow the water to go over the top. Water-saturated soil pushes against a wall, causing movement and bowing. Sight down the wall and use a level to determine the lean of the wall and to inspect it for **bowing**. Most walls are tilted back, so a plumb wall can be an indication of a developing problem.

Look for **drainage holes** toward the bottom of the wall and for evidence that the drainage holes are indeed working. Wood walls have natural openings, so weep holes won't be present.

Report **cracking** in poured concrete walls and suggest the owner patch the wall and monitor it for ongoing movement.

INSPECTING RETAINING WALLS
• Soil too low behind wall
• Poor drainage
• Tilting or bowing
• Cracking or deterioration

*Photo #21 shows a **wooden retaining wall** with a considerable bow to it. This wall had been inspected in the past and showed evidence of distress. It had been built to lean back, but at the time of the inspection it was already plumb. The customer was warned that problems were developing with the wall. We suggested then that tie-backs be added. On a whim, we returned three years later to see how the wall was doing and found it in this condition. The customer hadn't taken our advice. Instead, the owner is now trying to brace the wall, but without success. The bow in the wall tells the story. This movement won't stop now. The wall needs to be completely redone.*

#21 Wooden retaining wall

Reporting Your Findings

When you arrive at the house, fill in some general information about the structure. It's a good idea to record the **approximate age** of the house (you may want to ask the owner or Realtor for this information). You may want to state the style of the home and which direction the front of the house faces.

We also suggest that you record **weather conditions** on the day of the inspection. Reporting whether the house is snow covered, whether it's raining or dry, and the temperature is always a good idea. It helps explain why you might not be able to get on the roof (snow covered or wet) or why the basement is not leaking (conditions dry).

During the exterior inspection, be sure to communicate with your customer continually as you inspect attached structures, the walks and driveway, the grading and vegetation, and the retaining walls and fencing. Talk about what you're inspecting and what you're finding. Here's an overview on how to record your findings of the exterior inspection:

- **Walks and driveway:** Identify the types of sidewalks present and state their general condition. Note if walks are pitched toward the house and if any **trip hazards** are present. Identify the type of driveway material and report on its condition. Again, note any trip hazards.

- **Patio:** Identify the patio material and report its condition. Don't miss reporting if the patio is sloped toward the house or has trip hazards.

- **Balcony:** Always be sure to record any **safety hazards** with balconies. Remember, any missing balusters or railings, any balusters too far apart, loose railings, rotted wood, or instability in the balcony constitutes a potential safety hazard. Note the general condition of the balcony.

- **Deck:** Note the deck materials and their general condition. You might want to record whether wood decks are treated, painted, or stained and comment on the condition of the finish.

- **Stoops and stairs:** Identify the materials used in construction and note the general condition of the stoops and stairs. Be sure not to miss any trip hazards.

- **Porch:** If there's a porch present on the home, identify the construction materials and state the general condition of the porch and note defects with the structure. Be sure to identify the type of support piers present, noting that they're *not visible* if you can't see them. Make a special note if the porch is missing a railing that should be present for safety.

- **Retaining wall:** Identify the type of retaining wall present and record any defects you find with it.

- **Landscaping affecting the foundation:** It's important to indicate the location (east, west, north, or south side of the house) of any reverse (negative) grading found during the inspection and to suggest additional backfill or window wells if that's a solution. The situation may be such that you should recommend a soil engineer be called to look at the grading.

- **Safety hazards:** When you find safety hazards during your exterior inspection, it's a good idea to record them both on the page for that part of the inspection and on a summary page at the back of your report. Don't miss them. Always pay attention to balusters too far apart, missing handrail, loose railings, and dangerous balcony conditions.

DON'T EVER MISS

- Missing handrails and railings
- Balusters too far apart
- Unstable railings and supports
- Trip hazards
- Rotted wood
- Reverse grading
- Poor retaining walls

WORKSHEET

Test yourself on the following questions.
Answers appear on page 58.

1. Which of the following is required to be inspected according to most standards of practice?

   A. Sidewalks
   B. Fencing
   C. Soil conditions
   D. Safety glazing

2. What is the acceptable depth of treads, according to most building codes?

   A. About 1"
   B. 7" to 8"
   C. 9" to 11"
   D. Any depth is allowed.

3. In general, when is a handrail required on steps?

   A. If the steps are wider than 3' across
   B. If the steps are covered with outdoor carpet
   C. If there are 3 or more risers
   D. If there are open risers

4. What would <u>not</u> be the cause of a sag in a porch ceiling?

   A. Warped beams
   B. Beams that have slipped off the support columns
   C. Leaking into the structure of the roof
   D. Rotting fascia board

5. Which statement is <u>false</u>?

   A. A deck is an independent structure.
   B. Wooden support posts on decks should not have their feet buried in the ground.
   C. There should be no space left between floor boards of a deck.
   D. Generally, codes require balusters on decks to be 4" apart.

6. What can be the cause of driveway upheaval?

   A. Extensive salting
   B. Tree roots
   C. Uneven soil settlement
   D. Acid spills

7. What is a deadman?

   A. The cantilevered beam or joist that supports a balcony
   B. The cross piece in a tie-back that anchors a retaining wall
   C. The vertical piles in a retaining wall
   D. The drainage holes in a retaining wall

8. The home inspector is required to:

   A. Determine if the landscaping on the property has a pleasing effect.
   B. Recommend window wells for homes that don't have them.
   C. Recommend a soil engineer to determine the pitch of the land near the foundation.
   D. Determine whether the land slopes toward or away from the foundation.

9. Which of the following should be reported as a safety hazard?

   A. Reverse grading
   B. Deck flooring not properly spaced
   C. Balusters too far apart
   D. Water penetration at the intersection of the balcony and the house

10. Which of the following statements is <u>false</u>?

    A. A retaining wall should hold back water.
    B. A retaining wall should hold back the soil.
    C. A retaining wall should allow water to pass over it.
    D. A retaining wall should allow water to pass through it.

Chapter Five

INSPECTING THE GARAGE

The final item we're going to study on exteriors is the garage. The complete inspection of the garage includes the following: Roofing and drainage system

- Siding and Trim
- Windows and doors
- Structure
- Floors
- Vehicle door and automatic opener
- Electrical

In this chapter we won't be concentrating on the roofing and drainage system or the electrical. These items are presented in other of our guides.

Siding and Trim

Garages may be separate structures, completely **detached** from the home. Others may be **attached**, sharing a common wall or walls with the living area of the house. Attached or interior garages may also sit underneath rooms built over them.

Inspect the **siding of the garage** with the same diligence that you did the house. Identify the type of siding present and report on its condition. Pay attention to the following as described on pages 5 to 28 in this guide:

- Deterioration of siding
- Joints and interfaces
- Loose or missing components
- Fastenings
- Warping, cracking, buckling, and detachment
- Caulking
- Water penetration

Next, inspect the **exterior trim** on the garage. Check out trim as you did with the house as described on pages 30 and 31. Be particularly careful when you examine a detached garage, which may have had less attention paid to it than the house. Sometimes, homeowners will paint the house more frequently than the detached garage, won't have caulked as often, or may

Pages 57 to 60 present procedures for the inspection of the garage. This inspection normally includes the garage roof and electrical work. These two aspects of the garage inspection are presented in other of our guides — A Practical Guide to Inspecting Roofs and A Practical Guide to Inspecting Electrical.

INSPECTING

THE GARAGE

- Siding and Trim

- Windows and doors

- Structure

- Floors

- Vehicle door and automatic opener

have replaced trim on the house but not the garage. When you inspect the garage trim, pay attention to the following:

- Wood rot
- Broken or split trim pieces
- Loose or missing pieces
- Loose nails
- Paint and caulking condition
- Proper trim installation

Windows and Doors

The home inspector should inspect windows and doors to the garage as stated in pages 32 to 36 in this guide. Be sure to operate all "people" doors and inspect them as you would other exterior doors. (We'll talk about the vehicle door starting on page 63.) Examine **doors** for the following:

- General condition
- Door operation
- Framing
- Caulking
- Header
- Level and plumb
- Any storm or screen doors present

The **passage door** from the garage into the house should be fire rated. Some local codes may require a solid-core wood or a steel clad door. And if the door has a window, it may have to have wire-reinforced glazing or be of a certain size (although generally prohibited). Another requirement is the level of the bottom of the door. In general, the garage floor should be 4" below the living space to allow gasoline vapors space to dilute. Your local codes will state the requirements necessary to prevent fire and vapors from the garage passing into the house through this door. Improper passage doors should be reported as a **safety hazard**.

Inspect **windows** from the exterior and the interior. Garage windows may be fixed-pane windows or an opening variety. Windows in the garage should be examined just as carefully as those on the house. At this time in the inspection, garage windows should also be opened and tested for smooth operation. In many cases, these windows are never used and may not work

Worksheet Answers (page 56)

1.	A
2.	C
3.	C
4.	D
5.	C
6.	B
7.	B
8.	D
9.	C
10.	A

very well at all. They may even be painted shut, which should be pointed out to the customer. Inspect all garage windows for the following:

- Framing
- Glazing
- Caulking
- Level and plumb
- Operation

Inspecting the Structure

Garages, particularly the structure of the garage, can be neglected from the start. Builders may supply less than adequate bracing. Often detached garages are built without foundations and footings with the wood frame sitting directly on the slab or ground. Homeowners often pay less attention to the condition of the garage than they do to the house. Don't make any assumptions about garage structure after viewing the house. Conditions may vary considerably.

When you begin the inspection of the garage, look it over **from the outside.** Pay attention to any evidence of racking, twisting, tilting, or sagging. Eye the line of the roof for evidence of structural failure.

Once on the inside of the garage, check the visible structure. Don't overlook the **ceiling joists**. Often, garage ceiling joists are only intended to prevent rafter spread and to keep the walls from spreading and are not intended to support storage. Homeowners often store equipment along the joists. The home inspector should be sure joists are strong enough to support the load.

Note the condition of the visible framing. Be sure to check at the foot of walls for a **rotted sill**. If wood framing is sitting on the slab or ground, the sill is likely to have some damage to it. Inspect the rest of the garage structure for the following:

- Warping, twisting, and sagging
- Cracked, rotted, and cut members
- Loose fastenings
- Water penetration
- Delamination of sheathing
- Proper supports in place and secured

Personal Note

"One of my inspectors was inspecting a 1-car detached garage and found the sill entirely rotted out. He was curious, I guess, to see how bad the problem was and gave the side of the garage a nudge. "He held his breath as the side of the garage tilted, lifting about 3" up at his side, hung there, and finally settled back down. A stiff wind could take that garage down. The inspector was extremely thankful that he wasn't the one to do it."

Roy Newcomer

Photos #22 and #23 show a garage interior. In Photo #22, notice the short beam at the right. That beam supports the above structures at the right of the photo. The beam is supported by the vertical steel column. Or is it? Photo #23 shows a close up of the top of that steel column. The plate at the top of the column should be secured to the beam. In this instance, however, instead of eight nails as there should be, there's only one nail and that's not even nailed in. All that's needed to bring the column, the beam, and the other framing down is for someone to drive into the garage and tap that column with the car. We found this situation during an inspection. You can be sure we reported it and wrote it up as a safety hazard.

#22 Garage interior

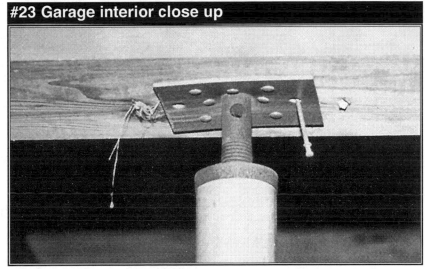

#23 Garage interior close up

SAFETY HAZARDS

- Absence of firewall

- Passage door not fire rated

- Exposed flammable insulation

- Garage floor less than 4" below basement or living space floor

- Heating or water heater unit less than 18" above garage floor

Distortion of the garage framing can be noticed by the fit of the doors, especially the vehicle door. If the framing is racking or twisting, the doors may not fit their frames and be hard to operate.

Safety Concerns

There are two main safety concerns with the garage. One is **fire resistance** and the other is protecting the living area of the house from **car exhaust fumes** and **gasoline vapors**.

Any surface of the garage that abuts the house — abutting walls and the garage ceiling — should be fire resistant and sealed against fumes. Wood frame walls between the garage and living area should have drywall on both sides of the studs with finished joints. If the wall is concrete block, there should be a vapor barrier on the inside. Of course, the passage door should be fire rated, solid-core wood, or metal, as mentioned on page 58.

#24 Proper firewall

Photo #24 shows the **proper firewall** in an attached garage. . When inspecting drywall, you'll want to be sure that joints are finished. Notice the proper height of the step-up to the living area from the garage in the photo.

These abutting walls and ceiling (if living space present above) should be **insulated** to the same degree as the external walls of the home. Insulation may not be visible to the home inspector, may be seen in holes or gaps in the wall facings, or may be exposed. Flammable insulation such as polystyrene and polyurethane foam should have a fire resistant cover. If the insulation is exposed and is combustible, it should reported as a **serious safety hazard**. Exposed combustible insulation is quite common in garages.

Gasoline leaking from the car poses a danger. Gasoline vapors are potentially explosive and can ignite easily. These vapors are heavier than air and tend to hug the floor. That's why the passage door from the garage to the house or basement should be **at least 4" above** the level of the garage floor. If vapors accumulate in the garage, they will be diluted enough for safety by the time they reach the 4" level. Where the heating equipment and/or water heater are in the garage, they should be at **least 18" from the floor**. Undiluted gasoline vapors could ignite when they hit the open flame of an oil or gas burner. At the recommended height, gasoline vapor would be diluted enough for safety. Report conditions less than those indicated as **safety hazards**.

The Floor

The home inspector first identifies the type of floor in the garage. It could be dirt, gravel, asphalt, or wood, but is most commonly a concrete slab.

The **concrete slab floor** in the garage can be laid too thin — it should be at least 3 1/2" thick — with or without steel reinforcement. Inspect the floor for the following defects:

- **Settling fill:** Concrete slabs are generally laid on grade on fill. Some may not have a layer of gravel for drainage or a moisture barrier. Later, the fill settles or washes away, causing the concrete slab to settle, twist, or fall. The home inspector can tap the floor in suspect areas. A hollow sound will indicate voids under the slab. If the floor has settled, there may be a line of concrete still adhering to the walls where the slab once sat.

- **Cracking:** Minor cracks in the slab are quite common and may not indicate any structural problem with the garage. But they should still be reported. Serious cracking is another story. Cracks that run from wall to wall may indicate that the foundation has settled or footings have failed. They should be investigated for causes. Cracks around sunken or hollow areas will indicate that the slab is settling.

- **Spalling:** The surface of the slab may crumble away. This can result from salts that penetrate the floor.

- **Drainage:** The garage floor should have a slope to it or a floor drain to allow water to flow away. The home inspector should check any garage drains for condition. Often, these drains are neglected and become plugged or broken. If the slab is settling in places, water may not be able to flow from the garage or towards the drain.

The home inspector may come across a situation where the **garage is built *over* a room.** In this case, the floor may be constructed of heavy timbers or of steel-reinforced concrete. The suspended wooden floor can rot. The suspended slab can experience rusting in its steel structural members and spalling of the concrete. The floor system may be weak, but this situation is very difficult to assess during the general home inspection. The home inspector should inform the customer of this and recommend that a **structural engineer** be called in to determine the condition of the suspended garage floor.

Vehicle Door

The vehicle door to the garage may be made of wood, aluminum, steel, Masonite®, or fiberglass. The home inspector should identify the door's materials and condition, noting broken panels, loose joists, deteriorating paint, unpainted surfaces, and so on.

Hinged doors: The home inspector should open these doors and test for smooth operation. With these heavy doors, rubbing can indicate a problem with the hinges, which should be strong enough and tightly secured. The doors may sag from their own weight unless properly braced to lift the outer edge up. When you open hinged doors, notice if there is the right amount of clearance for them to swing open freely. Do the doors bang into a wall of the house? Has vegetation grown up to restrict their movement?

Swinging doors: This type of door swings up to open. Lifted manually, the entire door swings overhead without "bending." It is assisted by a mechanical system of counterweights and/or springs. The door should be tested for smooth operation and its hardware inspected.

- **Manual overhead doors:** Overhead doors are made of hinged horizontal panels that "bend" along overhead tracks as they are opened. The home inspector should operate this type of door. The movement should be smooth. Manually operated overhead doors should hold their position when raised about 1/3 up. They should rise by themselves above that point and lower themselves below that point. It should be noted if the doors close with too much force. If movement is difficult, check for dented or twisted tracks, loose bolts, rusted rollers, and so on. Note if door panels are cracked or broken.

  NOTE: We often find overhead doors, both manual and automatic, painted on only one side. Usually, there's a manufacturer's sticker still of the inner side of the door saying that the inside should be painted. Be sure to point this out to your customer if the inside isn't painted. When the door has paint on only one face, it can warp and rot.

- **Automatic overhead doors:** Always test the operation of automatic doors. First, run them up and down while noting how freely they move. There should be no noise such as squeaks and no halting or jerking movements. Noise and halting indicates a maladjustment in the motorized mechanism or in the hardware.

The home inspector must check the automatic door for **safety**. There should be two **safety reverse features** on the door which will automatically reverse the movement of the door if it encounters an obstacle upon closing. This feature is especially important to families with small children. Do not attempt to stop the door with your hands. You can hurt your back, or fiberglass can break under stress. Place a block of wood at the door opening and close the door. Testing of the electric eye feature can be performed by blocking the beam. The door should reverse upon touching the block of wood or blocking of the beam. A malfunctioning safety reverse feature must be reported as a **safety hazard**.

As with the manually operated overhead door, check the automatic door itself and its hardware for condition. Again, you might find the inside of the door unpainted.

Reporting Your Findings

Be sure to continue your ongoing conversation with the customer as you inspect the garage. Although the garage is a relatively small structure, there are so many features of the garage to discuss — both from the exterior and inside the garage. Point out what you're inspecting and what your findings are. Be sure to discuss any safety hazards you find in the garage such as a missing firewall, dangers from fumes, a malfunctioning safety reverse feature on the automatic door, and so on.

Your inspection report should have a separate page or portion of a page for reporting on the garage. Don't skimp on reporting space for the garage. There's a lot to report. Here's an overview of what to report:

- **Garage information and structure:** Define the type of garage you've inspected with terms such as *attached*, *detached*, *2-car*, and so on. Note the condition of the garage structure, giving its overall structure a rating of satisfactory, marginal, or poor. Note defects such as racking, problems with ceiling joists, rotted sills, inadequate support columns, and so on. We suggest that you write a recommendation to call a structural engineer to evaluate a suspended slab when the garage is built over a room.

- **Fire issues:** Be sure to report on the presence or absence of a properly rated passage door to the house. Record whether or not the proper firewall separation is present. Make a note of the type of insulation present and identify exposed polystyrene or polyurethane insulation as a **safety hazard**.

- **Siding and trim:** As described earlier in this guide, record the type and condition of the garage's siding and trim. Pay attention to the caulking job on the garage and give the caulking a rating as to condition.

- **Vehicle door:** Identify the type of door present and report on its condition.

- **Automatic opener:** Indicate whether or not an automatic opener is present, whether or not it operates, and whether the safety reverses operate. Be sure to report a **safety hazard** if either safety reverse feature is not functioning.

- **Floors:** Describe the type of floor in the garage and write your findings on its condition. Make note of both typical cracking and large cracks that may be present.

- **Safety hazards:** For those safety hazards listed here, report them on the garage page of your report and repeat them again on a summary page. That way, customers can look over a single page to be reminded on any safety hazards found on the property.

— Improper passage doors to the house
— Passage door not 4" above garage floor
— Heating unit burners not 18" above garage floor
— Potential structural failure such as collapse
— Absence of firewall
— Exposed combustible insulation
— Safety reverses on automatic openers not operating

DON'T EVER MISS

- Rotting siding or trim
- Rotted sill
- Structural problems
- Missing firewall
- Passage door not fire rated
- Exposed flammable insulation
- Improper heights of passage door and mechanical equipment
- Floor settlement
- No safety reverse on vehicle door

WORKSHEET

Test yourself on the following questions.
Answers appear on page 68.

1. If the siding on the garage is the same as on the house, it does not need to be inspected.

   A. True
   B. False

2. The home inspector must test the garage door remote control transmitter to see if it works.

   A. True
   B. False

3. According to most standards of practice, the home inspector must operate all doors in the garage.

   A. True
   B. False

4. Which of the following statements is <u>false</u>?

   A. The passage door must be 4" above the garage floor.
   B. The passage door should be fire rated.
   C. The passage door must be 18" above the garage floor.
   D. Local codes may dictate whether the door should be solid-core wood or steel clad.

5. Which provides the proper fire resistance on the wall between the house and the garage?

   A. Polystyrene or polyurethane insulation with a fire resistant cover.
   B. Drywall on both sides of the wall studs with finished seams.
   C. Drywall on the inner side of the wall studs with a vapor barrier.
   D. Exposed flammable insulation.

6. Windows of the garage need only be inspected from the outside.

   A. True
   B. False

7. Why should heating equipment in the garage be raised off the floor?

   A. It protects the equipment from rusting.
   B. It keeps the heating equipment away from car exhaust fumes.
   C. It keeps the heating equipment from tipping if the floor settles.
   D. It keeps the heating equipment away from gasoline vapors.

8. A deep crack in the garage floor that runs from wall to wall may indicate:

   A. Spalling.
   B. Poor drainage.
   C. Foundation settlement.
   D. Salt erosion.

9. When might a structural engineer be recommended to inspect a garage floor?

   A. When there is a hollow sound made by tapping the floor
   B. When the garage floor is suspended over a room
   C. When the slab has settled
   D. When there are cracks around sunken areas

10. Which of the following should be reported as a safety hazard in the inspection report?

    A. Floor settlement
    B. No safety reverse feature on the vehicle door
    C. Fixed-pane windows in the garage
    D. Delamination of roof sheathing

11. If the garage's wood framework is sitting on the ground, the home inspector should:

    A. Suspect a rotted sill.
    B. Expect racking of the framework.

Worksheet Answers *(page 66)*

1.   B
2.   B
3.   A
4.   C
5.   B
6.   B
7.   D
8.   C
9.   B
10.   B
11.   A

EXAM

A Practical Guide to Inspecting Exteriors has covered a great
many details involved in the inspection of the exterior of a home.
Now's the time to test yourself and see how well you've learned
it. I included this exam in the guide so you'll have that chance,
and I hope you'll try it.

To receive Continuing Education Units:
Complete the following exam by filling in the answer sheet
found at the end of the exam. Return the answer sheet along with
a $50.00 check or credit card information to:

American Home Inspectors Training Institute
N19 W24075 Riverwood Dr., Suite 200
Waukesha, WI 53188

*Please indicate on the answer sheet which organization you are
seeking CEUs.*

It will be necessary to pass the exam with at least a 75% passing
grade in order to receive CEUs.

Roy Newcomer

Name_____   Phone:_____

Address_____   e-mail:_____

_____   Credit Card #:_____

                                           Exp Date:_____

*Fill in the corresponding box on the answer
sheet for each of the following questions.*

1. Which action is required by most standards of
   practice?

   A. Required to operate all windows on the
      home.
   B. Required to report whether the garage
      door remote control transmitter is
      working.
   C. Required to report whether the garage
      door operator will automatically reverse
      or stop when meeting resistance.
   D. Required to inspect all outbuildings.

2. Which of the following, according to most
   standards, must be inspected in the exterior
   inspection?

   A. Storm windows
   B. Garden sheds
   C. Awnings
   D. Wall cladding

3. What is the main objective of the inspection
   of exterior components of a property?

   A. To identify major deficiencies on the
      exterior
   B. To report exterior safety hazards
   C. To describe materials used on the exterior
   D. To determine the pitch of the grading

4. On which type of exterior wall structure would you <u>definitely not</u> find wall sheathing?

   A. Newer platform frame homes
   B. Older platform frame homes
   C. Balloon frame homes
   (D) Solid masonry homes

5. What is <u>not</u> an example of wall sheathing?

   A. Building paper
   B. Tongue and groove planking
   C. Plywood
   D. Fibrous sheets

6. If a wall is moisture impermeable that means:

   A. Moisture is allowed to pass through the wall.
   (B) Moisture will be trapped in the wall.
   C. Flashings are missing under the siding.
   D. The wall has insulation in it.

7. What is clapboard siding?

   A. Wood planks laid vertical with battens
   B. Wood planks laid vertical with scarfed joints
   (C) Wood planks laid horizontally and overlapped
   D. Wood planks laid horizontally in shiplap style

8. Nails that don't grip in vertical plank siding may be caused by:

   A. Deteriorating paint or finish stain.
   B. Nails that are too long.
   (C) The absence of wall sheathing.
   D. An improper nailing base.

9. Cupping and checking in wood plank siding is a sign of:

   A. Improper nailing
   (B) Moisture retention
   C. Movement of the frame
   D. Crooked wall studs

10. What condition should be investigated further in the clapboard siding as shown in Photo #1?

    A. The height of the lower course
    B. The corner moldings
    C. The placement of the downspout
    (D) The bowing at the lower courses

11. Which type of siding has the <u>least</u> ability to hold paint over long periods?

    A. Stucco
    B. Composition board
    (C) Aluminum
    D. Plywood

12. What is the difference between wood shingles and wood shakes?

    (A) Shingles are sawn; shakes are handsplit.
    B. Shingles are bigger.
    C. Shakes have to be double coursed; shingles don't.
    D. Shingles have to be painted; shakes don't.

13. Which siding becomes brittle in cold weather.

    A. Aluminum
    B. Steel
    C. Stucco
    (D) Vinyl

14. The home inspector should check overlaps where aluminum planks meet lengthwise to be sure:

    (A) They're big enough to prevent water penetration.
    B. That flashing is present.
    C. They're not rusting.

15. If tapping on a stucco wall causes a dull thud noise, the home inspector should suspect:

    A. Foundation settlement
    (B) Stucco detachment

16. Which condition might need to be reported as a safety hazard in your inspection report?

   A. Clogged weep holes in brick veneer wall
   B. Spalling brick
   C. Brick veneer detached from wall
   (D) A rusted lintel over a window

17. When the solvent in paint evaporates too quickly, what can be the result?

   A. Alligatoring
   (B) Chalking
   C. Crazing

18. The home owner can use bleach to test for:

   A. Rust stains on paint
   B. Chalking paint
   C. Mildew on paint
   D. Crazing paint

19. What is the fascia board?

   A. The portion of the roof that extends beyond the wall
   (B) The flat trim board fastened to the outer edge of the roof rafters.
   C. The underside of the eave
   D. The lowest course in wood siding

20. What obvious defect should be reported in Photo #3?

   A. Missing trim pieces
   B. Siding that needs to be replaced
   C. A rotted window sill
   D. A window without safety glazing

21 What is glazing?

   A. A gas such as Argon between lights
   B. A sticky plastic inner layer between lights to hold pieces of glass if the window breaks
   (C) The putty or compound used to hold glass in the sash
   D. The window pane itself

22. Identify the window components marked as A,B,C,D, in the drawing below:

   A. Upper sash, casing, sill, lower sash
   B. Casing, upper sash, lower sash, sill
   C. Casing, sill, upper sash, lower sash

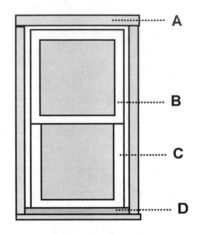

23. What would not be a cause of a slider door rubbing during operation?

   A. Worn tracks and bearings
   B. A leaking seal in the window
   C. A sagging lintel above the door
   D. Distortions of the door frame

24. Identify the stair components marked A,B,C,D, as shown in the drawing below:

   A. Tread, riser, run, stringer
   B. Riser, tread, nosing, stringer
   C. Riser, tread, run, stringer

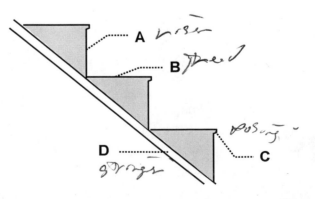

Exterior

1   C
2   D   Well Cladding.
3
4   A
5   D
6   B
7   C
8   C
9   B
10 D
11 C
12 A
13 D   vinyl
14 A
15 B   stucco detachment
16 C
17 B
18
19 B
20
21
22
23
24
25
26
27
28

25. What is an acceptable height of risers, according to most building codes?

    A. About 1"
    B. 7" to 8"
    C. 9" to 11"
    D. Any height is allowed.

26. What is the definition of a porch?

    A. A roofed extension of the house
    B. An independent platform attached to the house
    C. An above-ground platform protruding from the house
    D. A flat, paved area abutting the house

27. Which of the following statements is false?

    A. The home inspector should probe porch ceilings and fascias for wood rot.
    B. The home inspector should not inspect the porch.
    C. The home inspector should report missing railings on a porch over 2' from the ground.
    D. The home inspector should report a porch floor that tilts away from the house as a defect.

28. Why should deck floor boards be laid with space between them?

    A. To prevent cupping and checking
    B. To allow water to evaporate at the seams
    C. To eliminate the need for staining
    D. To provide water resistance

29. Which statement is true?

    A. Deck support posts can have their feet buried if the wood is pressure treated.
    B. A deck is built on the home's foundation.
    C. The home inspector should inspect the deck from underneath if possible.
    D. A deck is usually not fastened to the house.

30. What is usually the cause of a balcony that tilts downward when weight is applied?

    A. Rotted flooring members
    B. Loose railings
    C. Missing balusters
    D. Cantilevered joists that can't carry the load

31. What is the proper pitch for a patio?

    A. Perfectly flat
    B. Pitched away from the house
    C. Pitched toward the house
    D. It doesn't matter what the pitch is.

32. If there are offset cracks in a concrete patio, the home inspector should:

    A. Offer to come back and repair the cracks
    B. Suggest a structural engineer be called in to examine the cracks
    C. Report the cracks as a trip hazard
    D. Suggest outdoor carpeting to cover the cracks

33. What can be the cause of driveway pitting?

    A. Poor quality of materials
    B. Tree roots
    C. Uneven soil settlement
    D. Stones moving to the surface

34. What is the proper grading for land around the foundation?

    A. 1" per foot for 5" or 6"
    B. 1" per foot for 5' or 6'
    C. 5" per foot for 1' or 2'
    D. 5" per foot for 5' or 6'

35. Window wells might be recommended if:

    A. Reverse grading is corrected leaving the basement window below grade.
    B. Any basement window doesn't have one.
    C. The basement window is leaking.
    D. Shrubbery is too close to the house.

36. Identify the components marked A,B,C, in the wood retaining wall shown in the drawing below:

A. Anchor rebar, deadman, spikes

B. Gravel fill, deadman, spikes

C. Gravel fill, spikes, deadman

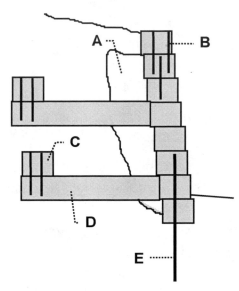

37. When should a retaining wall be inspected?

A. If bowing is noticed
B. If it is holding back soil near the house or garage
C. If there's evidence of water flowing through the wall
D. Every retaining wall should be inspected.

38. Even if garage trim is the same material as that on the house, it should be inspected.

A. True
B. False

39. Twisted aluminum siding on the garage may be an indication of:

A. Expansion and contraction
B. Missing J channels
C. Structural problems with the garage

40. **Case Study:** You are inspecting a 2-car garage attached to the house at its east wall, where drywall is present. The brick veneer siding shows step cracks on adjacent walls at the northwest corner of the garage. Inside the garage, there is an offset crack in the concrete floor at that same corner. The ceiling joists are sagging where some lumber is stored. A patio abutting the south wall of the garage is tilted toward the garage. Upon testing, the vehicle door won't reverse when you put a block of wood under it.

What, if any, condition would you report as a safety hazard in your inspection report?

A. Firewall missing
B. Step cracks in siding
C. Safety reverse on door not operating
D. None of the above

41. For the case study, what would you suspect is happening at the northwest corner?

A. Foundation settlement or footing failure
B. Veneer detached from walls
C. Rafter spread
D. Concrete slab not thick enough

42. For the case study, what suggestion should be made to the customer?

A. Replace veneer at northwest corner.
B. Patch the offset crack.
C. Remove the lumber stored in the ceiling.
D. Tell the kids to stay away out of the garage.

43. For the case study, what should be inspected carefully based on the pitch of the patio abutting the garage at the south wall?

A. The fascia boards on that wall
B. Bracing of studs in that wall
C. The sill along that wall

44. If there is a concrete block wall between the house and garage, there should be:

   A. A vapor barrier on the house side of the wall.
   B. A vapor barrier on the garage side of the wall.
   C. Drywall on the garage side of the wall.
   D. A stucco finish on garage side of the wall.

45. The passage door between the house and garage should:

   A. Not have a window.
   B. Be fire rated.
   C. Be more than 4" above the garage floor.
   D. All of the above

46. When inspecting garage structure, the home inspector should look for:

   A. Cracked, rotted, or cut framing members
   B. Warping, twisted, or sagging framing members
   C. Delamination of sheathing
   D. All of the above

47. When performing an exterior inspection, which of the following would you list as a safety hazard in your Inspection report?

   A. Rotted soffit, mildew on paint surfaces, flammable insulation in garage
   B. Corroded metal grills at window wells
   C. Negative grade, rotted soffits, rotting boards on porch stairs
   D. Missing railing on deck 4' above ground and rotted boards on porch stairs

48. Why should a home inspector record the weather conditions on the day of the inspection in the inspection report?

   A. If snow covered, to explain why the roof wasn't inspected
   B. If raining, to explain why the exterior inspection was not performed
   C. If dry, to explain why the trim wasn't inspected for leaking
   D. If very hot, to explain why doors to an air conditioned home weren't operated

49. When talking to the customer, the home inspector should:

   A. Make sure the customer understands the chemistry of paint.
   B. Only report good things about the house.
   C. Be sensitive to the customer's level of understanding.
   D. Make uneducated suggestions about repairs.

50. During the exterior inspection, the home inspector should:

   A. Inspect soil conditions.
   B. Inspect vegetation near the foundation.
   C. Inspect recreational facilities.
   D. Inspect for safety glazing

GLOSSARY

Alligatoring A process in which the solvent in paint evaporates too quickly, leaving residual solvent in the paint and causing wrinkling or cracking of the surface.

Aluminum siding Horizontal planks of aluminum with a baked-on enamel surface, used as a wall cladding.

Anchor post In a tie-back, the wooden post that connects to a wooden retaining wall and extends back into the soil.

Asbestos cement siding A mixture of portland cement and asbestos fibers in shingle form, used as a wall cladding.

Asphalt composition siding Asphalt-impregnated felted mats coated with an asphalt formation and covered with a granular material, used in shingle form as a wall cladding.

Awning window A window hinged at the top to open outward.

Balcony A platform protruding from the house that is not supported by the ground.

Balusters The vertical poles that support the railing of a staircase.

Battens Narrow strips of wood placed over joints in vertical wood plank siding to seal the joints.

Bay windows Three windows set at angles with each other in a bay that protrudes from the structure.

Beveled Clapboards that are tapered planks rather than perfectly rectangular in cut.

Bow windows More than three windows set at angles with each other in a bay that protrudes from the structure.

Box bay Bay windows where the windows are set in a bay at right angles to each other.

Brick ties Accordion-style metal fasteners used to attach a brick veneer to the wood framework of a house.

Brick veneer A wall construction method in which an outer layer of bricks is attached to the wood framework of the house using brick ties.

Building paper A paper installed between plank sheathing and wall cladding that acts as a water repellent and air barrier.

Casement window A window hinged at one side to open outward.

Casing A wood piece covering the edge of a window or door frame where it meets the wall cladding.

Caulking A waterproof material used to seal joints at interfaces between building components, used with some wall claddings.

Cavity wall A masonry wall with a dead air space left between inner and outer layers of masonry.

Chalking A process in which ultra violet radiation causes the vehicle in exterior house paint to break down and pigment particles to be released.

Checking In wood plank siding, a crack or split along the grain as a result of cupping.

Clapboard Overlapping, horizontal wood plank siding made from either rectangular planks or tapered planks.

Closed cornice Trim and moldings at the eave with both a vertical fascia board and a horizontal soffit.

Composition board Planks or sheets of compressed wood fibers with weather resistant binders, used as a wall cladding.

Compound wall A solid masonry wall built of two different materials.

Cornice The trim and moldings at the eave line.

Course Each row of siding material.

Crazing A condition in a paint surface where the new layer of paint shrinks while drying causing a net-like pattern of cross cracking.

Cupping In wood plank siding, a warp across the grain of the board.

Deadman In a tie-back, a cross piece spiked to the anchor post at the end in the soil, used to anchor a wooden retaining wall to the soil.

Deck An independent structure or platform that is attached to the house.

Detachment With veneer or stucco wall cladding, the separation of the siding material from its attachment to the house.

Double course An application of wood shingles or shakes where an undercourse, not exposed to the weather, is covered completely by a top course.

Double hung window A window with two sashes which both move.

Eave The overhang or lower portion of the roof that extends beyond the outer wall.

Fascia A flat trim board fastened to the outer edge of the roof rafters..

Firewall A fire resistant wall used where garage walls abut the house, preventing the spread of fire.

Fixed-pane window A window that does not open or close.

Flashing Sheet metal used at interfaces between building components to prevent water penetration.

Glazing The window pane made of glass or other material.

Grading The slope of the land around the house.

Head The top piece in a window frame.

Header A horizontal framing member that carries the load above a window or door opening. Also called a lintel.

Header rows Rows of bricks turned small end out to act as ties to hold a brick wall together.

Hopper window A window hinged at the bottom to open inward.

Jalousie window A window with narrow strips of glass that move together, lifting out from the bottom as the window opens.

Jambs The side pieces in a window frame.

J channel A manufactured component of an aluminum or vinyl siding system which has a curved channel that the planks fit into. Used around window and door openings to make a weathertight seal.

Joists Horizontal members of a floor system that carry the weight of the floor to the foundation, girders, or load-bearing walls.

Laminated glass A multiple light glazing with a sticky plastic inner light that holds broken pieces of glass together when the glass breaks.

Light Each layer of glass making up a window.

Mildew A fungus that can live on and in paint.

Moisture permeable A surface that allows moisture to pass through it.

Multi-pane window A window with small pieces of glass set into wood or lead muntins.

Multiple lights Glazing consisting of two or three layers of glass or other material.

Muntins A grid of cross pieces of wood or lead that hold small panes of glass in a multi-pane window.

Overhead door A garage vehicle door made of hinged panels that "bend" along overhead tracks as the door is opened. May be manually or automatically opened.

Patio A flat, paved area abutting the house.

Picture window A large fixed-pane window.

Pigment The color particles in paint.

Pitting Disintegration of the surface of the driveway material caused by poor quality materials or installation, acid spills, or the use of a de-icing salt.

Pivot window A window that pivots open from a center hinge.

Plywood siding Plywood sheets, some with a decorative or grooved outer surface, used as a wall cladding.

Polystyrene foam boards A plastic rigid board insulation used as a wall insulator. Must be covered for fire safety.

Polyurethane foam boards A plastic rigid board insulation used as a wall insulator. Must be covered for fire safety.

Porch A roofed extension of the house that is built as a part of the house.

Porch columns Vertical members that support the porch roof and the floor system.

Retaining wall A wall constructed to hold back soil.

Reverse grading A condition where the land around the house slopes toward the foundation.

Riser The vertical portion of a step in a stairway.

Safety glazing Glazing that is held in place when it breaks. May be tempered or laminated glass.

Safety reverse feature A mechanism on an automatic overhead door that will reverse the movement of the door if it encounters an obstacle when closing.

Sash The framework in a window that holds the glass or other material.

Scarfed joint A joint used in plywood siding where edges of abutting sheets are angle cut to fit snugly and prevent water penetration.

Sheathing See *Wall sheathing*.

Shiplap A style of milled plank used in plank siding that is laid close enough to appear to be butted.

Siding See *Wall cladding*.

Sill The bottom piece in a window frame. Also, the 2 x 4 or 2 x 6 laid flat and anchored to the foundation, providing a pad for the framing system.

Single course An application of wood shingles or shakes where each course is exposed to the weather.

Single hung window A window with two sashes, only one of which moves.

Slab floor A poured concrete floor, as in a garage, that rests directly on the ground.

Slider window A window with a sash that moves horizontally.

Soffit The horizontal board laid on the underside of the eave.

Solid brick wall A wall construction where three layers of brick are used to construct a solid wall with no wood framing.

Solvent The third constituent in paint that evaporates after the paint is applied. Also called the thinner.

Spalling The crumbling and falling away of the surface of bricks, blocks, or concrete.

Stop molding A bottom framing piece in a door frame that stops the movement of the door.

Stringer The side supporting member that supports a stairway.

Stucco A water resistant, plaster-like material made of sand, cement, and water, applied and used as a wall cladding. May have an acrylic finish.

Tempered glass Glass that shatters into small, smooth edged cubes when it breaks.

Thinner See *Solvent*.

Tie-back A heavy wooden post and cross piece used to anchor a wooden retaining wall to the soil behind it.

Tread The horizontal portion of a step on a stairway.

Trim The pieces added to the exterior siding that protect the framework from water penetration.

Upheaval A condition where sections of a driveway rise due to poor construction, an insufficient base, or tree roots and stones moving under the surface.

Vehicle The film-forming compound in paint.

Veneer See *Brick veneer*.

Vinyl siding Horizontal polyvinyl chloride planks, used as a wall cladding.

Wall cladding A siding or covering for the exterior of the house that protects the framework of the structure.

Wall sheathing Sheets of plywood or wood planking used to cover the wall framework of the structure.

Wall studs Vertical wall framing members.

Weep holes Openings in the bottom row of brick in a veneer wall providing an exit for water accumulating behind the veneer.

Window frame The framing that surrounds and holds the sash.

Wire mesh A mesh attached to the wall sheathing and studs used to anchor a stucco base coat to the wall.

Wood plank siding Rectangular wood planks, installed vertically or horizontally as a wall cladding.

Wood shakes Thick, rough, uneven shingles that are handsplit, split and sawn on one side, or sawn on both sides, used as a wall cladding.

Wood shingles Shingles that are sawn and are of uniform thickness, used as a wall cladding.

INDEX

A Practical Guide to Inspecting Program
Study Unit Two, Inspecting Exteriors

Student Name: _____ Date: _____

Address: _____

Phone: _____ Email: _____

Organization obtaining CEUs for: _____ Credit Card Info: _____

After you have completed the exam, mail *this exam answer page* to American Home Inspectors Training Institute. You may also fax in your answer sheet. You will be notified of your exam results.

Fill in the box(s) for the correct answer for each of the following questions:

1.	A□ B□ C□ D□		24.	A□ B□ C□		47.	A□ B□ C□ D□
2.	A□ B□ C□ D□		25.	A□ B□ C□ D□		48.	A□ B□ C□ D□
3.	A□ B□ C□ D□		26.	A□ B□ C□ D□		49.	A□ B□ C□ D□
4.	A□ B□ C□ D□		27.	A□ B□ C□ D□		50.	A□ B□ C□ D□
5.	A□ B□ C□ D□		28.	A□ B□ C□ D□
6.	A□ B□ C□ D□		29.	A□ B□ C□ D□
7.	A□ B□ C□ D□		30.	A□ B□ C□ D□
8.	A□ B□ C□ D□		31.	A□ B□ C□ D□
9.	A□ B□ C□ D□		32.	A□ B□ C□ D□
10.	A□ B□ C□ D□		33.	A□ B□ C□ D□
11.	A□ B□ C□ D□		34.	A□ B□ C□ D□
12.	A□ B□ C□ D□		35.	A□ B□ C□ D□
13.	A□ B□ C□ D□		36.	A□ B□ C□
14.	A□ B□ C□		37.	A□ B□ C□ D□
15.	A□ B□		38.	A□ B□
16.	A□ B□ C□ D□		39.	A□ B□ C□
17.	A□ B□ C□		40.	A□ B□ C□ D□
18.	A□ B□ C□ D□		41.	A□ B□ C□ D□
19.	A□ B□ C□ D□		42.	A□ B□ C□ D□
20.	A□ B□ C□ D□		43.	A□ B□ C□
21.	A□ B□ C□ D□		44.	A□ B□ C□ D□
22.	A□ B□ C□		45.	A□ B□ C□ D□
23.	A□ B□ C□ D□		46.	A□ B□ C□ D□